First published 1993 by Ashgate Publishing

Reissued 2018 by Routledge
2 Park Square, Milton Park, Abingdon, Oxon OX14 4RN
605 Third Avenue, New York, NY 10017

Routledge is an imprint of the Taylor & Francis Group, an informa business

A Library of Congress record exists under LC control number: 93172315

Electronically typeset using Microsoft Word 4 in 24pt, 18pt, 16pt, 14pt, 12pt and 9pt Palatino on an Apple Macintosh SE by Tony Bidgood

ISBN 13: 978-1-138-61206-8 (hbk)
ISBN 13: 978-0-429-45729-6 (ebk)

Contents

Contents

Preface

When we were writing the *CASE Handbook for Information Managers* (Avebury Technical, 1991), we received a considerable amount of information from various CASE vendors about the strategies and directions they were taking with their products. This led us to believe there was a need for a book which brought a number of the major vendor strategies together and reviewed them, within the context of a technical architecture to support the development of information systems for an organisation.

Vendors provide two different approaches to the way they describe the future of their products. There are what we call the 'visionaries' and the others. The visionaries produce a consistent framework within which all their products and, more often than not, other vendors' products fit together, whilst the others are content merely to produce a series of statements of intent for their product with more limited roles for third party vendors' products. With the move towards open systems becoming a major force within the IT industry, we have only concerned ourselves with reviewing the strategies of the visionaries.

These visionaries divide into two basic categories: hardware vendors, such as IBM and Digital, and software vendors, such as Computer Associates and Texas Instruments. In practice, these two software vendors typify two ends of a spectrum. The former are similar to the hardware vendors, slotting their considerable portfolio of software tools into a framework – a 'bottom up' assemblage of parts which they hope will make a whole. The latter has basically one product set which is being evolved 'top down' to take advantage of emerging standards and the move towards openness.

The book is divided into three sections. In the Introduction, we have set the scene by discussing the factors that determine the future of CASE and a review of the different types of CASE technology.

This section concludes with an examination of the 'punter's dilemma' – that of actually choosing a vendor strategy to use in earnest to provide the foundation for the development of information systems.

In the second section, we look at seven different strategies for the development environment from six hardware vendors: IBM (both AD/Cycle and AIX CASE), Digital, ICL, UNiSYS, Hewlett Packard and Bull.

In the third section we again review seven different strategies, this time from software vendors. Three of these have approaches broadly similar to the hardware vendors: Computer Associates (CA), Software AG and Manager Software Products (MSP). The remaining four are Oracle, Softlab, LBMS and Texas Instruments; each offers something rather different and distinctive, based around a single product or family of products.

There is an appendix on the concepts of Total Quality Management – increasingly important both to the providers and consumers of CASE.

We would like to thank the following people for their help in providing us with information about the strategy of their organisations:

- The IBM Consultants Desk
- Chris Martin and Dr. Sami Zahran of Digital Equipment Corporation
- The ICL INFORM Desk
- Bob Richards, Steve Clinton and Dr. Mike Hamm of UNiSYS Europe-Africa
- Ray Daniells of Hewlett Packard
- GJ McNeil of Bull HN Information Systems Ltd
- Robert Hailstone of Computer Associates
- Nigel Martin and Dr. Wolfgang Martin of Software AG
- Richard Barker and John Spires of Oracle
- Roger Learmonth and Elise Rabelais of LBMS
- Eric Magnussen and Gary Jenkins of JMA Information Engineering Ltd (part of Texas Instruments)
- John Grieve of Softlab UK
- Phil Redfern of Manager Software Products

This book can be read on its own as a high level guide to the strategies of the certain vendors. It can also be read in conjunction with the authors' *CASE Handbook for Information Managers*, where a reader is actually involved in the evaluation process and wants to obtain some independent information on the future directions of particular vendors in the CASE market.

We would like to thank John Hindley of Ashgate (formerly Avebury Technical) for all his help and encouragement during the gestation period of this book.

And most importantly ...

Our final thanks go to our wives, Sue and Ann, and our children, David and Sarah, and Amy, David and Lewis, for putting up with yet another year of us spending hours locked away in our studies writing this book.

Simon Holloway
Berkhamsted
June 1992

Tony Bidgood
Bewdley
June 1992

Section A

Introduction

Section

Introduction

Chapter 1

What Determines the Future of CASE?

1 Introduction

The future of CASE – computer aided systems engineering – will be determined by a battle which has been raging for a decade. As with all conflicts, a variety of competing forces is engaged in combat. These forces consist of technical opportunities and constraints, vendors' and users' business objectives and critical success factors, and intellectually inspired advances in methodologies and techniques. Generally, the factors involved can be categorised as demand and supply: what the market wants and what the vendors can produce.

Demand factors include the following:

- developments in methods and techniques
- business imperatives
- maturity of the IS/IT profession.

Supply factors include the following:

- emerging technical standards
- vendor interests.

1.1 Developments in methods and techniques

Recent developments in methods and techniques have placed conflicting demands on the suppliers of CASE. On the one hand is the continuing formalisation of methodologies, illustrated by:

- the efforts of EC governments to define a 'Euromethod' for systems development
- the move to make Version 4 of SSADM a British standard for analysis and design
- certificates of proficiency awarded by examination for SSADM and for Information Engineering
- the integration of Information Engineering methodologies and toolsets such as IEF and ADW.

On the other hand is the increasing urgency of systems development. The familiar one-to-three–year cycle for developing and implementing a new system has been rejected by organisations who want results in months, weeks or even days. The IS/IT response has been varied but, in some cases, dramatic. The widespread adoption of Rapid Application Development (RAD) methods, techniques and tools has allowed, simultaneously, a cut in timescales and improvements in quality.

These trends do, though, have something in common: the recognition that quality, and the dramatic improvement of quality, is vital to the success of IS/IT functions and of the organisations they serve. CASE has a critical rôle to play in this crusade: it helps to ensure that two things happen, without which a quality product cannot be delivered.

- CASE facilitates communication:
  - between developers, users and business sponsors
  - between different project stages, from strategic planning through to implementation and beyond.
- CASE populates the 'project database' with all the necessary information about application(s) under development or in production.

Only where methodologies and techniques have been implemented rigorously and effectively, using the best available tools, can success be achieved, and the increasing sophistication and complexity of business applications reinforces this statement.

During the last five years, a number of radical software products, allied to appropriate methodologies such as RIPP from DuPont (marketed in the UK as Corvision from Cortex) and the RAD and IE-RAD variants of IEM from James Martin and Co., have allowed unheard-of productivity gains without loss of quality. Two very different examples are discussed briefly – AD/Advantage from

CINCOM and Sapiens from SAPIENS. Both rely extensively on expert systems technology, and this element is crucial to their success.

- AD/Advantage provides a high level of support for reuse and component management, and incorporates sophisticated change control facilities. It focuses on facilitating RAD via prototyping and screen report painting, utilising a set of 'integrated software robots' to aid the developer in producing a system.

  AD/Advantage imports analysis objects from conventional front end tools (Excelerator, ADW, ECT, IDEF/Leverage and others), and uses these specifications 'to automatically generate design objects and reusable process components'.[1] These reusable components can be automatically converted into code (COBOL or other options) for various platforms, with 'executable panel designs and file views ... dynamically bound to an AD/Advantage program at run time'. They can then be changed, using CUA-compliant SQL-based editors, at any time without changing the program or the process.

  This provides a level of integration at the back end which promises great savings of time and effort, while ensuring that quality objectives are met.

- Sapiens is a rules-based, object-based development environment. This allows working models – applications – to be developed rapidly on the basis of facts and business rules, processed via the expert systems knowledge base within the product. Development is iterative, and new models can reuse or replace components from other, earlier, models.

  Using Sapiens, a UK-based petrochemical engineering company redeveloped all its core systems and developed many new systems. Since 1987, no project has taken longer than three months from initial requirements definition to production, and the development team size is typically four people, including users throughout.

Neither of these products removes the need for business analysis, but the detailed requirements definition process typical of conventional systems development is pared back. For instance, the engineering company quoted above uses a simplified subset of SSADM to define the inputs that Sapiens needs, ignoring process modelling and producing mainly data definitions.

1.2 Business imperatives

In the 1990s, the most successful businesses will be those which can respond most rapidly to the market. Organisations are looking for ways to cut timescales and costs – in many cases by 80% or more. Time based competition, business process redesign, total quality management and component based manufacture are vital elements.

1.2.1 Time based competition

Time based competition, whereby faster product development and more rapid delivery of products and services provide competitive edge, is driving many companies to demand the same response from IS/IT. Neal Monnery, of the Boston Consulting Group, states that organisations should 'make time consumption the critical management and strategic parameter' and 'use increased variety and enhanced responsiveness to stay close to their customers'.[2] He quoted typical improvements in production flow times (Figure 1.1), based on BCG client assignments and interviews.

Monnery identifies four strategic implications:

- as time is compressed, productivity increases
- as time is compressed, prices can be increased
- as time is compressed, share increases
- as time is compressed, risks are reduced.

1.2.2 Business process redesign

Such time compressions can be achieved by cutting out steps, by reordering processes, and by carrying out activities in parallel. Working with suppliers and customers can really help: a major car manufacturer recently saved 80% on its purchase order costs by cutting out the whole invoicing process: goods are supplied, and suppliers paid. There is no extraneous paperwork, and any discrepancies are handled as exceptions.

The key phrase here is business process redesign, described by Davenport and Short[3] as 'the analysis and design of work flows and processes within and between organisations'. The organisations that succeed in the 1990s will be those which can use IT to drive changes in the way they operate, achieving 'IT-enabled redesign'.

	Before	After	Percent Reduction
United States			
Motorcycles	360 days	<3 days	99%
Motor controllers	56 days	7 days	88%
Electric component	24 days	1 day	96%
Europe			
Twist drills	21 days	3 days	86%
Gun reloaders	42 days	3 days	93%
Japan			
Washing machines	15 days	<1 day	99%

Figure 1.1 Typical Improvements in Production Flow Times (Monnery)

Davenport and Short identify CASE as one of the key facilitating technologies: 'The ability to draw models rapidly and make changes suggested by process owners speeds redesign and facilitates owner buy-in. ... A further benefit is that when the business process changes, the IS organisation can rapidly modify the affected system.'

1.2.3 Total quality management (TQM)

All organisations and professionals want to be quality performers. Yet quality is a major problem for industry as a whole, and for the IT sector. In Japan, many companies have successfully used total quality management (TQM) to turn themselves around or deal with corporate crisis. There is a widespread awareness that quality leads to productivity and profitability, and quality and productivity can be seen as two side of the same coin.[4]

The three fundamental components of TQM are unit optimisation, horizontal integration and vertical alignment.

- Unit optimisation requires all individuals and groups of within an organisation to stabilise and improve their own work, reducing variability and providing a baseline for later improvement efforts.
- Horizontal integration demands that all departments and project teams concentrate on satisfying their users together, across organisational boundaries.
- Vertical alignment requires that all managers and professionals understand and contribute to the few crucial goals of an organisation.

TQM, and other structured approaches to achieving quality, are now being adopted throughout the world; they are emphasizing an already-apparent demand for information systems which have two key characteristics:

- they can be proved to conform to precisely-defined business requirements
- they are developed by teams using methodologies and tools which allow them to attain specific quality objectives.

These objectives cannot be achieved using the old hand-crafted approaches to software production, and require the use of an integrated CASE toolset to achieve the required results. Whether this level of integration is achieved by adopting a single life cycle product (IEF, ADW, or a very few others) or by following a more open 'framework' approach (AD/Cycle, Cohesion, etc.), which allows various tools to work together, is less important than the principle that software production must be 'industrialised' if it is to meet the quality objectives of the 1990s and beyond.

See Appendix A for a more detailed discussion of TQM.

1.2.4 Component based manufacture

As Malcolm Fowles has pointed out,[5] there is a worldwide pattern, seen in all successful industries, of a move away from individual hand crafting towards component based manufacture. There is striking evidence that IS/IT is also moving in this direction – for instance the US-based company which provides warehousing systems based on rapid customisation of an object-oriented generic warehousing application – and the CASE industry is, it hopes, poised to play a central rôle in this revolution. By creating and maintaining sets of business and technical objects which can be combined, adapted and recombined as required, users can at last begin to industrialise the process of software production and delivery. Figure 1.2 shows a component-based manufacturing model of software production – radically different from the traditional life cycle.

One key element is the increasingly important rôle of back end CASE tools. First generation CASE was almost exclusively a front end phenomenon (see Chapter 2 for the authors' categorisation of CASE tools), with products such as Excelerator and Automate – later Automate Plus – assisting the analyst to produce high quality data and function models of individual business systems and their users' requirements. In parallel with this was the culmination of a trend, apparent since the 1970s, towards 'front loading' the effort involved in the systems development process – analysis and design became ever more significant in the systems development life cycle. In fact, many projects never got beyond this stage, and the phrase 'paralysis by analysis' was coined.

To break out of this problem, prototyping seemed like the answer; why spend years analysing and designing when a working model of a system can be constructed in days or weeks, and then altered later if necessary? Without the right tools, however, this approach all too often proved unworkable. Prototyping in COBOL, for instance, does not fit the bill; it still takes a long time between iterations, and can never generate the right applications without rigorous specification. The key phrase here is 'a faster disaster'.

The use of rapid development tools, allied to the intelligent use of expert systems technology, has begun to attack the 'paralysis or disaster' dilemma, and products such as AD/Advantage and Sapiens, discussed above (section 1.1) are allowing organisations to create and reuse components in an industrial rather than a craft context.

1.3 Maturity of the IS/IT profession

This profession is a newcomer. In 1960, almost nobody was aware that computers were about to change the working lives of millions.

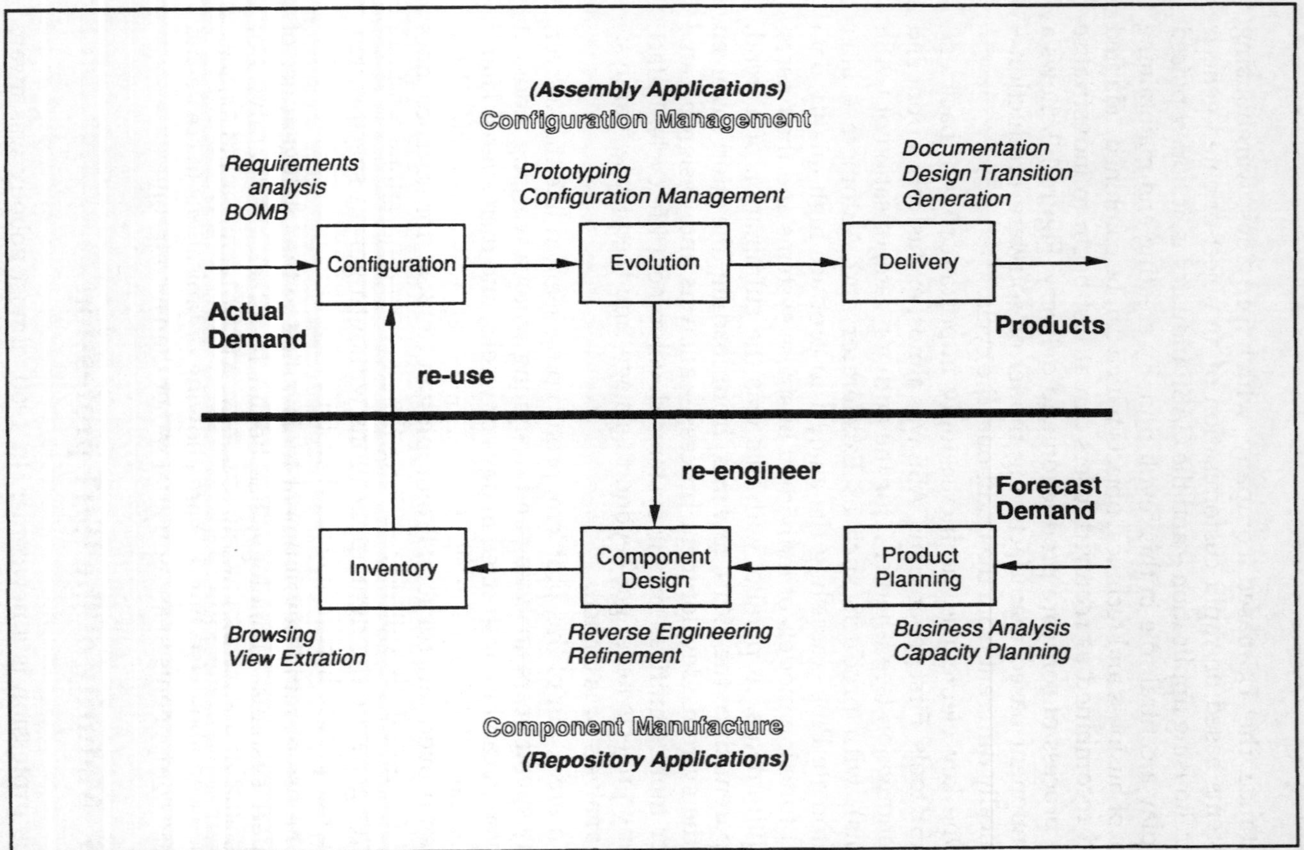

Figure 1.2 Component Based Manufacturing Model for Software (after M. Fowles)

In 1970, most large companies and government departments had one or two mainframes, and local authorities were just starting to invest. Data management was a new idea. The first data modelling projects had yet to be started.

In 1980, first generation (batch oriented, hard coded) systems were everywhere, databases were the big news, data dictionaries were gaining a foothold in the marketplace and a small proportion of developers had started to look to structured methods.

In 1990, organisations were struggling to keep pace with business demands as they kept their 1970s systems afloat, and many seasoned professionals had spent their working life within the IS/IT industry. Many had gained valuable experience of structured methods and of first generation CASE tools, and had started to become involved in the first, tentative implementations of information management programmes – building on earlier attempts to introduce data management – and the development of modular, reusable system components. Second generation CASE was now vital, though many organisations had made no significant investment in the first generation.

The challenge for the '90s and beyond is to harness the expertise and knowledge locked up in those systems and people, while refusing to allow them to become constraints on action. The people side of this is important, and in all the industrialised countries there are professional bodies and training schemes – public or private – aimed at keeping up the momentum for change.

There are some clouds on the horizon. It is a commonplace nowadays that we need to automate business, but the equal need to automate ourselves sometimes proves difficult to sell, as people see their set of skills, acquired over many painstaking years, becoming redundant. A change of mindset and a considerable training programme is essential if professionals – such as COBOL programmers, systems analysts, operators, dBASE whizzkids and the like – are not to be overtaken by events.

1.4 Emerging technical standards

The IT industry has been noted for its ambivalence to standards. Most IT professionals agree, if pressed, that standardisation is essential if their organisation is to improve its productivity and quality performance. Equally, the industry as a whole has been noted for its reluctance to commit to specific, agreed standards.

But significant progress has occurred. Fifteen years ago the idea of open systems, where an application based on (say) Digital hardware could easily share information with ICL or IBM-based systems, was seen by many as unrealistic. In the 1990s a single application could run on any of these platforms – or on all of them at once in a client server environment.

A similar trend has occurred within the CASE world. Early front end tools were stand-alone products, their output consisting of paper-based diagrams plus printed definitions and descriptions. This output could be used as the basis for program specifications and file definitions. Efforts were then made to automate this process, and file transfer mechanisms were created which allowed information to be passed between (for instance) Excelerator and Telon. But the process was never straightforward; where different hardware or software was involved, or where different import/export protocols had been adopted, success was not guaranteed.

Open systems standards are becoming a reality in the CASE arena: the closed architecture is, if not dead, fighting for its survival. First the Information Engineering I-CASE toolsets, led by IEF and IEW/ADW, showed the productivity and quality gains to be achieved by automation throughout the development life cycle. Then the fightback started: component CASE had to compete if it was to survive and it did this initially, as indicated above, by a series of one-off links between tools. But neither of these approaches was perfect. The I-CASE route restricted its adherents to a very small range of products and technical environments – primarily DB2/MVS – while the linked-component-CASE route was never comprehensive and only partially automated.

1.4.1 Open CASE

The challenge was (and still is!) to achieve an all-embracing framework into which all the necessary tools and methods could be slotted. The idea of open CASE is not new, having its foundations in the IPSE concept.

The architectural statements of IBM, Digital and, to a lesser extent, UNiSYS for the development environment appeared to have turned back the open CASE clock, by reinvigorating the idea of component CASE. However, roll back the covers of these architectural offerings and the joins are visible. The success of Digital's COHESION, relative to IBM's AD/Cycle is largely due to its use of *de facto* and *de jure* standards as the glue for those joins. The standards involved are many and varied, but there are four key areas of interest.

1.4.2 CDIF (CASE data interchange format)

CDIF is used for exchanging information between CASE tools and repositories. Information that can be exchanged includes descriptions, placement and details of text and graphic elements (such as process logic, entity relationship diagrams and DFDs, and including line shapes and font types and sizes). The standard has been developed under the auspices of America's Electronic Industries Association, and is to be adopted by ANSI.

1.4.3 Semantic transfer language

This is used for tool-to-tool communications. The standard can represent a composite set of CASE tool information with text and generalised graphical information. It is used as an intermediary language to express a powerful set of tool-generated information among different products in a conceptualised non-graphical form. The standard has been developed by the IEEE.

1.4.4 IRDS (information resource dictionary system)

IRDS defines what should be contained in a repository, but leaves it to the vendor to produce the technical implementation. It uses a data model to describe the way information is logically stored in the repository and the methods to be used by tools to access the information. At present the data model varies between the ANSI and ISO variations of the standard.

1.4.5 PCTE (portable common tools environment)

PCTE permits the use of CASE tools across different operating systems and LANs. It includes a common object-oriented communications layer for transparent information exchange between different types of products. It is primarily used for C, C++ and ADA environments. The standard was developed by the European Computer Manufacturers' Association (ECMA).

1.5 Vendor interests

Any organisation looking for a CASE strategy for the future must be aware of the pressures that vendors themselves are under. Vendors have, in the past five years, woken up very rapidly to reality. All of them, without exception, now appreciate two key facts of life:

- they cannot produce all the solutions themselves; and
- industry does not want to buy from a single source.

For certain of the large vendors, on both hardware and software side, this realisation came as a considerable shock, but in the '90s their strategies involve others' tools also. 'One stop shopping' is no longer available, though many of the vendors will be happy to direct the purchaser towards products that fit into their strategies.

1.5.1 Customer focus

The hardware vendors are, in a number of cases, downgrading the sale of processors, disk drives and workstations in favour of software. For instance, companies such as UNiSYS, ICL, Data General and Bull are rapidly moving towards an open systems strategy, providing upgrade paths for customers of their proprietary architectures. To keep these customers, they major on service, on systems integration, on package support and other features.

There is a general awareness that the future does not lie in box-shifting , and that software and systems integration is the way of the future.

1.5.2 Package implementation

For application developers, the future, or a significant element of it, lies with the implementation of packages, many of which can be tailored. For instance Customer 1 (Andersen) provides a generic template design, documented on a CASE tool, for a customer-handling system. This can be customised fairly rapidly for specific organisations, allowing faster implementation and – the assumption must be – higher quality software, given that a high proportion of the design is generic and thus proven elsewhere.

Such initiatives are a significant growth area, and many are produced by organisations other than CASE vendors. There are several known examples in the energy sector, for instance in terms of electricity supply, where several major utilities from two continents have been working on an industry template with worldwide sales potential. Without I-CASE the job would not be cost-effective, and customers would be relatively uninterested in buying a paper-based design.

1.5.3 Consultancy sales

Vendors have realised that consultancy can generate a considerable amount of revenue. CASE is a complex issue, and requires a significant degree of expertise to get the best out of it. Whether it relates directly to tool use, to the methodologies surrounding the use of CASE, or to wider issues such as business direction and re-engineering, consultancy may be seen by the vendor not only as a source of direct income, but also as a sales-generating activity.

Tom Peters refers repeatedly to the value to be gained by integrating your business with that of your suppliers and/or customers.[6] If you can be seen as an essential business partner, with a symbiotic relationship, then you can sell more, and the business will have a high degree of persistence – it is as simple as that. So the CASE vendors want to help you as much as they can, and it is up to you to control the situation.

2 Why focus on vendor intentions?

Since late 1989 we have seen all the hardware and software vendors put forward strategies for their products in the CASE arena. This book will look at the major strategies, based on the latest available information at the time of writing. The reader should, if interested in a particular strategy, contact the vendor for further developments.

Each announcement has a number of similarities, based on:

- the vendor's or hardware vendor's repository
- integration of their own products

- adoption of an 'open' strategy
- enhancing the image of their products in the marketplace.

The effect, particularly in the case of the major hardware vendors – primarily IBM, Digital and UNiSYS – has been to legitimise CASE technology for those who have not already seen the benefits. Fourteen CASE strategies, from thirteen vendors, are reviewed in this book:

Hardware vendors and their strategies

- IBM – SAA and AD/Cycle
  – AIX CASE
- Digital – NAS and COHESION
- ICL – CASE Partnerships and OPEN*framework*
- UNiSYS – Integrated Information Architecture
- Hewlett Packard – NewWave and CASEdge
- Bull – Distributed Development Environment

Software vendors and their strategies

- Computer Associates – CA90s
- Software AG – Integrated Software Architecture and ENTIRE
- ORACLE – Open CASE Environment
- MSP – RDADE
- Softlab – Maestro II and E-CASE
- LBMS – Engineer Framework
- Texas Instruments – Information Engineering Facility

Vendors typically have a clearer picture of where they want to go than do most users of CASE. After all it is their core business, whereas for most users the use of these tools is merely a means to an end, and not a primary part of their business value chain.

One example of a vendor with a clear picture of the future is Softlab (see Chapter 17), whose Chairman, Dr. Klaus Neugeberger, outlines the need for a combination of IPSE and I-CASE which he calls Enterprise-CASE, or E-CASE.[7] Softlab's Maestro II product incorporates what he identifies as the important features of an IPSE – tightly integrated life cycle tools for project and configuration management, with multiple architecture targets – and of I-CASE – front end and back end tool integration: 'For historical and cultural reasons, the I-CASE concepts will continue to thrive in the U.S. The IPSE concepts will continue to do well in Europe. Those users and vendors who can incorporate the two the quickest will benefit early from the emerging E-CASE movement of the 1990s.'

Here is a clear statement of market direction, coupled with a stated ability to satisfy consumer needs. It is notable for its focusing of the reader's attention on a relatively few, clearly stated, objectives. In this it is not unusual. Most, if not all, CASE vendors can produce statements explaining why their particular vision of the future, linked (naturally) to their own product sets, will satisfy market needs into the future. The problem for potential users of the products is that each of these visions is compelling, yet each is different.

Each vendor, in this increasingly crowded market sector, needs its own USP – its 'unique selling point' – in order to keep the public interested. Each is jockeying for survival, and for market share.

Some vendors attempt, paradoxically, to lock users into their own open systems world by stressing, though not actually providing, freedom of choice. 'Yes, you can transfer information between these products at any time' (but you will lose some of the meaning, the metamodels differ, and the interface is one way only). 'Yes, you can develop for any major platform using this tool' (but not the screen handling, the client server elements, or the database accesses).

Others rely on momentous announcements (e.g. AD/Cycle) some of which are actually backed by momentous action (e.g. the Object Management Group agreements which have paved the way, in a technical sense, for the much-heralded OO future). Such announcements should always be viewed critically: what is the actual substance, and what is the track record of the vendor or organisation? Caveat emptor!

Looking back at the AD/Cycle announcement in 1989, we see that it contained a lot of good ideas and intentions, but that the subsequent delivery has been poor. So the market is looking at competitors – both those in the IBM world, such as Computer Associates (CA), and those very definitely outside it, such as Digital (DEC) – to fill the gaps.

Having realised that they cannot go it alone, the vendors are continually manœuvering for position, forming alliances, pulling out of agreements, merging with rivals, and so forth. This is standard business practice, but there are few businesses where it happens with such bewildering speed. In the 1980s, IBM and Microsoft were regarded as friends. In the 1990s they are enemies. The reverse has happened with IBM and Apple. Sage (vendors of the back end CASE tool APS) and Excelerator (a front end CASE company) have combined to form Intersolv, which positioned itself as a life cycle CASE vendor within the AD/Cycle context. Intersolv itself, though, has its disagreements with IBM at the time of writing, but appears fundamentally committed to AD/Cycle.

This type of uncertainty all affects the user in the long run, and obviously causes considerable short term anxiety. The punter cannot afford to blow with the wind: organisations must commit to a strategy for a minimum of three years, and there are ways to interpret the issues.

For instance, a recent letter from Ian Palmer[8] exemplified the complexities involved in analysing the suitability of one vendor strategy – IBM's AD/Cycle – but suggested a way forward:

> The road to AD/Cycle so far has been tortuous and poorly signposted. Yet I would suggest that, given time, the road will be longer and better defined than those other architectural routes, initially mapped out in some marketing department of Big Blue. Eventually AD/Cycle will become as much a modern motorway as IBM's SNA and SSA.
>
> When AD/Cycle was first launched it was little more than a framework in which to position the by then established Information Engineering, case and expert system technologies, an area of integration that had been completely neglected by IBM. The Repository was, and remains, the keystone of the AD/Cycle architecture.
>
> Even so a repository with an information model designed by a committee of competing software vendors was inevitably going to be at best a camel and at worst a failure. Today the only viable solution for progressive IS organisations is to make use of a third-party repository such as Texas

Instruments' encyclopædia or Softlab's object management system.

Yet IBM cannot simply abandon AD/Cycle. It has little option but to provide a viable Repository. This requires rigorous definition of the objects in the information model, if data sharing between different tools is to be meaningful. But this in turn requires standardising on methodology concepts and techniques – so far anathema to IBM.

For the next couple of years, regardless of AD/Cycle, users with a serious commitment to case would be best advised to go for one of the major I-Case products. As AD/Cycle becomes a reality the I-Case players will inevitably provide a close coupling between their repositories and IBM's.

3 What does it mean for CASE users?

Figure 1.3 shows some typical features of CASE tools today, and compares them with facilities which we expect to be routinely available in the late 1990s. Note the use of the word 'typical'; most of tomorrow's facilities are available somewhere today, but not as a coherent set. Examples of the future today include the following, among many others:

- limited client server facilities exist in LBMS' System Engineer and Texas Instruments' IEF
- a degree of repository integration is displayed by MSP's ManagerView and by Digital's COHESION
- useful levels of version control and change control are provided by Andersen's Foundation and CGI's PacBase
- support for tasks which could not be carried out manually is provided by, among others, KnowledgeWare's ADW (comprehensive consistency and completeness checking)
- expert systems technology has been used by Sapiens (formalising and interpreting business rules) and by James Martin and Co.'s IE-Expert and AD-Expert (interpreting methodology rules and guiding developers)
- reusable components have been facilitated by CINCOM's AD/Advantage and by Aspect's LANSA.

TYPICAL CASE TODAY	TYPICAL CASE TOMORROW
SINGLE USER OR MULTIPLE SINGLE USER	TRUE MULTI-USER
STAND ALONE OR LIMITED LAN FACILITIES	FULLY DISTRIBUTED FOR DATA AND PROCESS VIA CLIENT SERVER ARCHITECTURE, LAN, WAN, etc.
PROJECT (SINGLE SYSTEM) ORIENTED	ENTERPRISE (MULTI-SYSTEM) ORIENTED
SINGLE TARGET PLATFORM OR NO TARGET PLATFORM	MULTIPLE TARGET PLATFORMS
CLOSED ARCHITECTURE	OPEN ARCHITECTURE
OWN HUMAN COMPUTER INTERFACE (HCI) OR SINGLE STANDARD GRAPHICAL USER INTERFACE (GUI)	MULTIPLE STANDARD GRAPHICAL USER INTERFACE (GUI)
LIMITED AND CUMBERSOME FACILITIES FOR WORD PROCESSING, DTP, GRAPHICS, DATABASE, DATA DICTIONARY LINKS, PROTOTYPING, etc.	FACILITIES EXTENDED AS REQUIRED BY INTEGRATION WITH OTHER TOOLS (USER TRANSPARENT)
INTERNAL DICTIONARY	MULTIPLE REPOSITORY INTEGRATION
LITTLE OR NO REAL SECURITY, VERSION CONTROL, CHANGE CONTROL, etc.	FULL CONFIGURATION MANAGEMENT
MANUAL EFFORT REQUIRED TO TRANSFER INFORMATION BETWEEN TOOLS	INFORMATION FULLY EXCHANGEABLE BETWEEN TOOLS
PARTIAL CONFIGURABILITY OR NONE	FEW CONFIGURABILITY RESTRICTIONS
PART LIFE CYCLE COVERAGE ONLY	FULL LIFE CYCLE COVERAGE BY INTEGRATION OF TOOLS
METHODOLOGY SUPPORT SPECIFIC TO SINGLE METHODOLOGY OR NO REAL METHODOLOGY SUPPORT	TAILORABLE (OR PRECONFIGURED) TO SUPPORT MULTIPLE METHODOLOGIES
AUTOMATES TRADITIONAL MANUAL TASKS	EXECUTES TASKS WHICH COULD NOT BE CARRIED OUT MANUALLY
CAN ONLY KNOW WHAT IT HAS BEEN TOLD EXPLICITLY	EXPERT SYSTEMS TECHNOLOGY TO SUGGEST DE FACTO BUSINESS RULES, APPROPRIATE DESIGN FEATURES, etc.

Figure 1.3 CASE Today and Tomorrow

The key phoneme in the 'CASE tomorrow' column is 'multi'. The products which succeed will be those which provide the best fit to different business and technical environments, whether by providing options as standard – support for different methodologies, graphical user interfaces, target platforms and so forth – or by becoming so open that all of these options are available by seamless links between tools.

The future is there for all to see – repository/configuration management, true multi-user environment, customer involvement, component management, client server technology and applications, open systems, multi media – but are *you* moving in this direction? is your CASE supplier?

What it means for the CASE user can be summed up very briefly by mixing a couple of metaphors:

- look before you leap, and
- don't sit on the tracks – there's a train coming!

Choose a vendor strategy, not necessarily one of those discussed in this book, which is right for your organisation today and which will be able to provide you with an upgrade path for the forseeable future. What are your organisation's objectives? What is your current position? How can you move from present to future. The TQM movement talks about the PDCA (plan–do–check–act) cycle of continuous improvement, discussed in Appendix A. You have to move, and you have to move now. But as Ian Palmer's letter, quoted above, makes clear, you do not have to make the big move immediately. There may be some short-to-medium-term action which gives real benefits with reduced risk, while paving the way for the future.

One last thing – whatever you do, you cannot sit back and let the future wash over you.

4 References

1 CINCOM (undated), *AD/Advantage – A Life Cycle Application Development System* (publicity booklet)

2 Monnery, Neal (1990), *Principles of Time Based Competition*, from seminar on Time Based Competition, OTR Group, London, June 1990

Chapter 2

Categories of CASE

Note: The contents of this chapter are extracted and adapted from Chapter 1 of the authors' 1991 publication *CASE Handbook for Information Managers,*[1] which contains more detail about the various categories, and lists all the (then) known tools in each category in its Appendix A. We have not repeated these lists here, but have given a couple of examples of each category for illustration.

---

1 Introduction

The 1980s saw a major software revolution take place, in the form of tools to help speed up the systems development life cycle. These tools, known generically as 'CASE tools' have taken many forms and tackled different parts of this task to greater or lesser degrees.

So, what is CASE? It is an acronym, standing for computer aided systems (or software) engineering.

Richard Barker[2] defined CASE as the combination of graphical, dictionary, generator, project management and other software tools that assist computer development staff to engineer and maintain high-quality systems for their end-users, within a framework of a structured method.

CASE tools in all their forms are changing the way organisations, and their IS/IT functions in particular, develop, support and maintain business systems. The tools have promised – and even delivered – improvements in efficiency, productivity and quality.

These are important issues: we have been forced to respond, and to respond ever more quickly, to rapid changes in the business environment.

However, there is another side to the picture. Much hype has been given to these products and, in a number of well-publicised reports, the hype has been exposed. So CASE is not some miraculous breakthrough in software technology which will solve all our problems; nor is it the panacea which many product brochures suggest.

Most products, of course, offer something worth having. Tim Bourne[3] also suggests that, to some extent, there is little new about CASE, because the ideas have already been expressed in the development of data dictionaries; rather, it is the combination of the well-tried ideas and the new technology that has added a new dimension to what is possible.

2 CASE categories: the pigeon holes

CASE has the capability to be all things to all men. There are almost as many definitions as there are CASE products. In this respect, Neil Holloway[4] proposed three basic product categories:

- **front end CASE** (or upper CASE or analyst workbench) tools, which support the belief that the need is for a productivity aid to help design and document systems, leaving the programming function to the language most suited to the environment
- **back end CASE** (or lower CASE, system generator or 4th generation system) tools, which support the belief that true productivity gains occur when the systems design is, automatically, converted into efficient code in such a manner that it may be easily maintained
- **life cycle CASE** (or I-CASE) products, which address both of the above activities, and are heavily tied into James Martin's masterplan for Information Engineering – the life cycle approach to software design and construction is the ultimate dream for many software developers.

This categorisation was developed further by Simon Holloway and Richard Williams,[5] who introduced two more groupings:

- **integrated project support environment** (IPSE) tools, which offer an environment which enables a system to be developed under a unified hardware and software platform
- **information resource planning** (IRP) or strategic planning tools, which provide support for the planning and high-level control of the information resource. They provide facilities which allow the information to be developed in graphical formats, and also support the use of concepts such as critical success factors.

In gathering material for the *CASE Handbook for Information Managers*,[1] the authors identified a further four categories:

- **reverse engineering** tools, which cover a number of approaches to the analysis of existing systems and to the reconstruction of existing code to improve both form and execution efficiency
- **project management** tools, which cover all aspects of project management, from support for the production of a plan, through resource utilisation to estimation
- **verification, validation and testing** (VV&T) tools, which cover all those aspects of software engineering which seek to ensure that a product has been developed soundly, performs as expected and is suitable for its purpose
- **configuration management** tools, which cover the management of versions and variants, the building of systems from constituent parts, and the operation of applying changes with suitable authorisation to established states through which a project passes.

As with so many types of software – word processors with DTP functions, spreadsheets which can produce presentation-quality graphics, drawing packages with search facilities – the distinctions between the different types of tool are becoming increasingly blurred, and many products could be differently pigeon-holed. However, the categiories remain valid, insofar as they provide a way for potential customers and users to view the marketplace and anlyse their needs. In the following sections we will look a little more closely at these nine categories.

2.1 Front end CASE tools

A spate of analyst workbenches has appeared since 1985. These tools support business analysts in documenting business requirements, and support the latest methodologies for systems analysis. The methods, and hence the products, rely heavily on graphics and, typically, these tools provide a graphical interface to allow an analyst to draw diagrams representing the results of his or her work.

Different kinds of diagrams can be produced, according to the methods supported by the vendor. Data models (or entity-relationship diagrams) are the most commonly provided, according to Rosemary Rock-Evans,[6] and data flow diagrams (DFDs) are almost as widespread. Various kinds of activity model, other than DFDs, are also common.

In addition, most products have rudimentary repository facilities, with attractive pop-up screen formats to fill in. Often these formats are closely linked to the diagram. Reporting facilities are in general fairly limited at present. There are many tools on the market; examples are Excelerator (Intersolv) and Design Aid (CGI in the US and Hoskyns in the UK).

The purpose of a front end CASE tool is to define the system requirements and the properties a system must possess to meet those requirements. Typical outputs of a front end tool are:

- system specifications, consisting of system components and the interfaces connecting them
- screen definitions
- report definitions
- data structures
- databases and files
- functional components.

Typical facilities of a front end CASE tool are:

- diagram editor, supporting at least three types of diagram (typically data models, DFDs and activity decompositions)
- screen and report painter
- dialogue prototyper
- repository with reporting capabilities
- export/import capabilities between external repositories and the product's repository.

2.2 Back end CASE tools

Back end CASE products, including those referred to as 4GLs, COBOL generators and other application generators, take as their input the results of analysis. From this they automatically or manually design screens, reports, logic and database structures with which a working, complete application can be generated. These tools use built-in design rules, reducing the amount of time and the level of knowledge that the application developer needs and improving the consistency of output for the user.

Back end tools have the potential to alter completely the process conventionally used to build commercial applications. Such products are increasingly able to take business data models and apply rules to design database structures for a specified database. Models of business activities can, still with a degree of manual intervention, be used to generate application logic in COBOL or other languages. In addition, generation across different hardware platforms is becoming more common, opening up the possibility of a new level of hardware and operating system independence.

Many application development tools on the market can replicate some or all of the functionality of COBOL, and promise major productivity gains for application development. Such products as Cincom's MANTIS, Software AG's Natural and DEC's RALLY have made a major impact on development costs when used in the appropriate way with modern techniques and organisation structures. However, they do not tackle the issues critical to the systems analyst, and their total contribution to solving the application development problem is thus limited.

Some of these products now incorporate front end CASE facilities: ICL, for example, has added DesignMaster on to the front of QuickBuild. There are more products available in this category than in any other; a couple of examples, from a very long list, are Telon (CA) and Focus (Information Builders).

Typical outputs from a back end CASE tool are:

- source and object code
- screen maps
- database and file definitions
- Job Control Language (JCL).

Typical features of a back end CASE tool are:

- screen painter
- report painter or writer
- prototyping support
- data modelling and repository support
- procedural logic handling and database access
- code and documentation generation
- specification manager to store program specifications
- testing and debugging tools
- export/import capabilities between the product and external repositories and databases.

2.3 Life cycle CASE tools (or I-CASE)

Although their early development was inspired by the Information Engineering philosopy, life cycle tools can be viewed simply as the combination of front end and back end CASE tools in one product. As might be expected, the outputs and facilities of these products combine those of front end and back end tools. Life cycle tools are sometimes referred to as I-CASE – Integrated **CASE** – as they provide integrated support for a given systems development life cycle. There are relatively few of these products on the market; the best known are IEF (Texas Instruments) and IEW/ADW (Knowledgeware), both of which support James Martin and Co.'s. Information Engineering Methodology (IEM).

2.4 Information resource planning (IRP) tools

These tools facilitate the process of ensuring that information systems development supports corporate objectives. The necessity for this has been increasingly realised over the past decade, and various tools have been developed, particularly by methodology vendors who need to support their own particular brand of information resource planning. Examples of this type of CASE tool are Entellect (Entellect) and Tetrarch/1 (PA Consulting).

Typical outputs from IRP tools are:

- high level data models
- organisation structure models

- association matrices
- data, function, application and technical architectures
- IS/IT plans
- reports.

Typical facilities of IRP tools are:

- data and activity diagramming, including data subject area diagrams
- repository
- report specification
- matrix manipulation
- critical success factors (CSF)
- Boston boxes
- application portfolio definition
- export/import facilities from the internal repository to front end CASE repositories.

2.5 Reverse engineering tools

Reverse engineering is the exact opposite of engineering. It aims to go back from code and files to the original system design and thence to the system requirement. Reverse engineering tools must extract details of the essential business functions and data from ageing, but critical, applications. This allows an organisation to salvage the investment of money and time, programming and database design skills, and user knowledge that built the original system. In essence, reverse engineering is a bottom-up process that has to assess the value and quality of existing systems by means of portfolio analysis.

The late Russell Jones[7] identified three sub-categories of reverse engineering tools:

- software resource analysis
- code restructuring
- true reverse engineering.

Software resource analysis tools, such as PM/SS (Adpac), document the content (structure, references, logic) of existing code in a structured fashion, allowing it to be replaced or rewritten based on a full understanding of the impact of change.

The code restructuring process is based on a number of key, well-accepted software engineering principles:

- structured programming constructs (sequence, iteration, condition)
- the theorem and the mathematics behind structured programming
- the method for converting unstructured programs to structured ones.

Code restructuring tools analyse the basic structure of a program, reorganise it into standard constructs without altering its logical consistency, and regenerate code, including as many of the original program's 'landmarks' as possible. An example of a code restructuring tool is Retrofit (XA Systems).

True reverse engineering tools are the future; the hope is that the next generation of products will be capable of abstracting the business logic essence of a program or database structure direct from code or data definition language (DDL). The abstract logic will be placed in a central repository, where developers can meld it with design information gleaned during the normal, forward engineering process of analysis and design. The whole thing will then be used as the basis for a new application. Currently very few products exist which go some way towards the achievement of this goal. An example is the Bachman toolset (Bachman).

2.6 Integrated project support environment (IPSE) tools

IPSE products originally derived from software project management needs, and were based around software specification and project control methods. Typically, IPSEs provide the sort of project and configuration management facilities that most current CASE tools still lack. There are, fundamentally, two types of IPSE: those with an open architecture and those which are closed.

At the logical centre of an IPSE sits the kernel. This basic set of facilities supports the base concepts recognised by the IPSE, and provides the means of building other concepts from the basic set. Further facilities may be supplied by the vendor which extend those of the kernel and are built using the kernel facilities. The user can use both the kernel facilities and the extended facilities, and these together constitute the infrastructure.

Current IPSE products provide facilities in a number of areas:

- user interaction, allowing varied and flexible support for menu, mouse and command line
- data management, supplying tools for object and meta-object definition to allow multiple versions of all project and system data to be recorded, accessed and combined as required
- activity management, providing the means to define process structure, tool invocation and method-specific process models
- presentation to the user by command inputs or procedure calls, because tool builders need to be able to access kernel facilities from within tools
- distributed systems support
- tool integration facilities, both those specially written for the IPSE and, if the IPSE has an open architecture, those provided by other existing tools.

Typical IPSE facilities are:

- word processing
- interface to electronic mail
- project management
- configuration management for version and access control
- project planning
- calendar and task assignment system
- estimator tool
- timetables and scheduling tool
- quality control metrics
- repository with systems audit trail.

Examples of IPSE products are ISTAR (Imperial Software) and Maestro II (Softlab). IPSEs are a peculiarly European concept, and the term is little known in North America, though Softlab is making a determined effort to break into the US market with Maestro II, badged as an E-CASE (enterprise **CASE**) tool – see Chapter 17.

2.7 Project management tools

The degree to which CASE tools are available to support individual project management functions varies considerably. Tasks such as estimating and scheduling are well supported but others, such as risk management or the handling of contractual conditions, have little or no CASE help at present.

The IT-STARTS Developers' Guide[8] describes project management as consisting a set of primary tasks:

- project justification, estimating and definition
- project organisation
- planning, progress monitoring and control
- configuration management and change control
- quality control and risk management.

The guide also identifies other areas, including tool and toolset management.

A more complete definition would include the various aspects of resource management, both of people (team building, leadership, motivation, etc.) and of physical resources (plant, equipment, software, etc.).

The first project management tools were to support the task of planning, and this is still the best supported area. Products run on the complete spectrum of hardware platforms, from PCs to mainframes.

Examples of project management tools are PMW (Hoskyns in the UK) and MS Project (Microsoft).

2.8 Verification, validation and testing (VV&T) tools

Verification, validation and testing cover the processes associated with ensuring that a product – any project deliverable – is developed correctly in the manner required, and that it meets its defined requirements.

Verification compares the output of each phase of the systems development life cycle with the requirements derived from the previous phase. The objective is to ensure that the deliverables produced by a phase fulfil all the requirements for that phase.

Validation checks that the original phase specifications were meaningful and appropriate, both in terms of the ultimate user requirements and in terms of the development methodology employed. In the later stages of system development, this may be achieved by executing a working prototype and observing its behaviour.

Testing is the process used to validate and verify.

Whilst VV&T applies to all phases of the system development life cycle, each phase requires different VV&T methods and techniques. Most of those developed for the early phases are manual, but the picture is changing with the introduction of front end and IRP tools with their built-in consistency checkers. By far the greatest number of VV&T methods and tools have been developed for the construction and implementation phases of the systems development life cycle.

Various tool classifications have been proposed, and a good summary is given in Pressman.[9] The only published book, by DeMillo et al.,[10] lists 218 products. In the USA, other sources of tool information are available.[11] The best classification the authors have encountered, and the one used in this book, is from Dorothy Graham.[12]

Graham classifies VV&T tools in the following categories, based on the type of facilities offered:

- capture/playback, to record and replay a test input script
- comparator, to highlight the differences between two files or programs
- input generation, particularly to support stress testing
- static analysis verification, to find errors without dynamically executing software
- simulator, to simulate some aspect of the real world in order to test software which could not be tested directly for whatever reason
- test harness/test driver, to run tests automatically,
- test coverage measurement, to assess how much of the structure of a program has been tested
- debugger, to identify and isolate software errors.

Graham found that these classifications could be used for general guidance but that, in many cases, an individual product's classification is less clear-cut. Examples of VV&T tools include a debugging tool – Abend-Aid (CompuWare) – and a combined capture/playback comparator and test harness – Evaluator (QA training).

2.9 Configuration management tools

We are increasingly faced with the need to monitor multiple projects. This means controlling all the items which contribute to the results of those projects. This activity is called configuration management.

A 'configuration', or 'working group', is simply a collection of *versions* of resource items. This concept encompasses a variety of collections required, during the development process, to describe and control the storage, sharing and use of resource items. In particular, this refers to particular 'releases' of software used in production.

Configuration management may be described as the process of:

- identifying and defining the configuration items
- controlling the release and the changes of these items throughout the system development life cycle
- recording and reporting the status of items, including change requests
- verifying the completeness/correctness of configuration items.

Configuration management consists of three elements:

- configuration control
- change control
- version control.

Together, they provide a system whose purpose is to ensure the availability, consistency and accuracy of the configuration items. A comprehensive configuration management system controls software, documentation and hardware items. It will assist the project manager to assess the state of a project by evaluating the formal status of its constituent parts: the configuration items.

Configuration management facilities are often found in other tools, including:

- source library management systems
- data dictionaries
- verification, validation and testing (VV&T) tools.

Two examples of configuration management tools are PVCS (Intersolv) and CCC (Softool).

3 References

1 Holloway, Simon and Bidgood, Tony (1991), *CASE Handbook for Information Managers*, Avebury

2 Barker, Richard (1989), *CASE*METHOD: Entity Relationship Modelling* , Addison-Wesley

3 Bourne, Tim (1990), *CASE Tools and the Future* , in *Using CASE Tools in Systems Development: Their Scope and Value,* edited by Williams, R., Gower Technical

4 Holloway, Neil (1987), *Trick or Treat,* ICP Business Review, Nov/Dec 1987

5 Holloway, Simon and Williams, Richard (1990), *Introduction,* in *Using CASE Tools in Systems Development: Their Scope and Value,* edited by Williams, R., Gower Technical

6 Rock-Evans, R. and Engelein, B. (1989) *Analysis Techniques for CASE: A Detailed Evaluation, vol. 1,* Ovum

7 Jones, Russell (1990), *Review of Current CASE Technologies,* in *The Distributed Development Environment: The Art of Using CASE,* edited by Holloway, S.R., Chapman and Hall

8 National Computing Centre (1989), *IT-STARTS Developers' Guide,* NCC

9 Pressman, Roger (1987), *Software Engineering: A Practitioner's Approach* , McGraw-Hill

10 DeMillo, R., McCraken, M. and Passafiume, J. (1987), *Software Testing and Evaluation,* Benjamin Cummings

11 *Testing Tool References Guide* and *Test Guide Update* from Software Quality Engineering

12 Graham, Dorothy (1990), *Software Verification and Testing Tools: Availability and Uptake* , in *SE90,* edited by Hall, P., BCS Conference Series, Cambridge University Press

Chapter 3

The Punter's Dilemma: Choosing a Vendor Strategy

1 Introduction

We are reminded of John Bunyan's words from Letters:

> And through the clouds I saw a shining palace with marble hall and towers topped in paradise ... as the sun shone through, now I saw a meaner palace with bars and spikes and walls; no palace but a prison.

IS/IT management should not start by looking for the most appropriate vendor's strategy. Technical strategies, of which the CASE strategy is just a part, need to be driven by business requirements. The authors would advise organisations to plan carefully, and not to put the cart before the horse. A suggested sequence is:

- determine business requirements, policies and constraints
- evaluate the current technical environment in terms of hardware, software, netware and methods
- determine IS/IT technical requirements for the future, based on the business requirements
- specify requirements for the IS/IT infrastructure (including organisational changes)
- select appropriate vendor(s) to support these requirements
- produce an implementation plan to manage the change process.

The selection of an appropriate vendor strategy must be considered when looking at the IS/IT infrastructure. It is important, however, to examine and understand each of the six stages, and this chapter summarises each of these, in order to put vendor strategy selection into its proper perspective.

2 Business requirements definition

From an IS/IT perspective, business requirements definition provides the key external driver of change. Three major questions must be answered when one analysing the organisation's need for IS/IT.

- What are its information needs?
- How well does the existing IS/IT department meet these needs?
- What does the IS/IT department require if it is to meet the currently unfulfilled needs?

These three questions are, in fact, the output from an information strategy planning (ISP) study. The major external input to an ISP is the corporate business strategy.

Having gained an understanding of the organisation's objectives and environment, the current status of the business, its possible future directions and how transitions are to be achieved are documented. The strengths, weaknesses, opportunities and threats are identified.

Potential key techniques identified by the CCTA (the UK government's Central Computing and Telecommunications Agency) include:

- Arkush square – to help in assessing the style of investment likely to be required
- Boston square – to help identify the rate at which demand for a product is growing and its likely profitability
- business systems planning – resulting in the production of a data architecture
- business modelling – to identify the high level entities and functions which are of particular importance to the business;
- critical success factors (CSFs) – the areas in which the organisation must achieve results in order to be successful

- PEST analysis – to document an understanding of the organisation's environment viewed from political, economic, social and technical viewpoints
- portfolio analysis – to provide a model against which requirements for existing and new information system can be assessed
- Porter's value chain analysis – to analyse an organisation in terms of the activities it performs, describing a linked set of primary and secondary (support) functions
- stakeholder analysis – to identify the major external influences on the organisation, their likely perceptions of the organisation, and their potential interactions with it
- SWOT analysis – to identify the internal and external strengths and weaknesses, opportunities and threats which affect the organisation
- user needs survey – to assess the real systems needs of users from a business perspective.

3 Current technical environment

The other key – internal – driver of change is derived from an analysis of current IS/IT performance and capability. This involves putting together an inventory of hardware, software, networks, applications, methods, techniques and organisation structures. Software, methods and techniques must be understood in terms of how, when, and by whom they are used.

The effectiveness and efficiency of the IS/IT inventory must be assessed, not only from an IS/IT perspective but also from a business perspective. Examine the current applications to see if they meet both current and future needs – in part? fully?

Identify and document the current development life cycle, both in theory and in practice – they may not be the same at all. What tasks are actually carried out, and what deliverables do they produce? Which current tools support which tasks and deliverables?

The existing IS/IT strategy (if one exists) must be mapped onto the identified business needs of the organisation:

- to identify progress against the strategy
- to identify scope for improvement.

Systems under development and those planned for the future must also be considered.

The extent to which the use of information systems has permeated the company should be analysed in terms of:

- the types of technology used
- the extent to which key business processes are supported by technology-based information systems
- the degree to which working practices are supported by systems
- the degree of integration between computer systems.

The current technical infrastructure needs to be looked at to ensure that it is really supporting current needs, and its degree of support for future requirements defined. Areas to be reviewed include:

- telecommunication systems – both service and data
- common computer systems
- office systems – text, data and graphics
- document storage and retrieved mechanisms.

Potential key techniques include:

- business impact analysis – this compares the extent to which current information systems support meets current business needs, and the extent to which it is likely to meet future business needs
- McKenny square – this assesses the importance to the business of information systems both now and in the future.

4 IS/IT future requirements

The third view looks at where IS/IT want to be in the future, and is driven by an understanding of the business requirements and the current IS/IT environment. The results should provide a strategy for managing the organisation's information resource, an architecture of the organisation's information, and a blueprint for meeting future business information needs.

The technical issues should include:

- standards relating to hardware, software and netware
- development methodologies and associated support tools
- service management methods and tools.

An IS/IT department may wish to change from being reactive to business needs, to being proactive. This will change the way that department must work with the business user community.

Business requirements for information systems may require support for particular categories of hardware, software, netware and 'methodware'.

In another situation, an IS/IT department might have identified a weakness or threat in their current infrastructure, and therefore have a need to counter this by adopting new features or strengthening existing ones.

5 IS/IT infrastructure

Based on the results from the previous three sections, IS/IT should look at how it can achieve its objectives, and support those of the organisation. Various types of question must be answered, and a few examples follow.

- What hardware platforms are appropriate for the future?
- What software is needed?
- What methods and techniques will be used to develop applications of all types?
- How should IS/IT be organised?

At this stage, it is appropriate to look at the architectural statements of the hardware and software vendors to see which best fit your requirements. So it is now that IBM's SAA, Digital's NAS, HP's NewWave, and so on, should be examined. And it is at this point that organisations must answer the question 'AD/Cycle or not?'. We would point out that every major vendor has announced some strategy. Organisations should not be afraid to have more than one hardware/software solution. In fact, in the 1990s, open systems are all about exploiting the vendors to gain the best for your organisation.

6 Implementation

The final stage is that of implementation. Here, the question 'What have I got to do?' must be answered. This means the construction of a plan covering all the steps needed to go from where you are now to where you want to be. The plan must include a time and responsibility element. In addition, the risks associated with the change must be documented, as also must be the (probably greater) risks associated with *not* changing. The steps in the plan will depend on where you are, what you have and where you want to go. We would expect some or all of the steps below to be included, depending on the current infrastructure:

- tool implementation
- methods implementation
- tool and/or method migration
- re-organisation
- recruitment
- training/retraining.

It is always important to remember, in today's quality conscious world, that garbage in still means garbage out.

7 Conclusion

All the strategies discussed in this book are the strategic directions taken by particular vendors. At the moment, the strategies of the big two – IBM's AD/Cycle and Digital's COHESION – carry implications for other vendors, and for the users of their products. This is borne out by the UNiSYS and Computer Associates architectures, each of which reference IBM and Digital. Each of the strategies announced is at a different stage of definition, evaluation and implementation.

The risks in adopting the wrong strategy include:

- a strategy which may become a blind alley
- inadequate functionality and structural limitations
- environmental complexity, interfacing and performance,
- lack of skills and adequate support
- disruption of service to the business by the introduction of new technology.

Do the benefits justify the costs? Any real benefits will not be apparent for a long time.

Organisations must understand the real issues that face them as they try to exploit their information resource. This means that they must ask the right questions. They must identify in detail their information requirements, and analyse how well their IS/IT departments are meeting the information objectives with their current environment of hardware, software and people.

If there is a need to change and evolve, then the organisation's requirements must be fully defined before looking at vendors' strategies: these are, after all, only solutions. The implementation process will need planning and managing – it will be the largest IS/IT project ever. However, we do not want to detract from the vision that vendor strategies, such as AD/Cycle and COHESION, have given us: they do provide the way ahead for most organisations, and certainly promise to provide the necessary development environment for the 1990s.

Section B

Hardware Vendors' Strategies

Chapter 4

The Hardware Vendors

Compared with the choice of software vendors featured in section C, the selection of hardware vendors covered in this book was relatively simple. We have featured most of the the 'traditional' mainframe vendors, those with a large installed base and a variety of architectures to support. These are the vendors who have needed, without exception, to present a coherent CASE strategy to the marketplace. Some of them have been able to achieve this to grater effect than others.

IBM, the largest supplier of computer hardware and system software, has a special place in the industry. In certain markets, such as Denmark, it is virtually the only mainframe supplier of significance. Chapters 5 and 6 explore the *two* CASE strategies marketed by IBM – AD/Cycle and AIX CASE. The fact that these two strategies exist is testimony to the diversity and depth of IBM's penetration of world IT, but it also suggests at least an element of the right hand not knowing what the left hand is doing! This last may partially explain IBM's (by now) legendary habit of promising what it cannot deliver.

Chapter 7 describes, by contrast, the unified and consistent approach taken by Digital (DEC). This company's NAS and COHESION strategy is based upon a functioning and well-proven repository, and integrates all Digital platforms into the same environment.

Chapters 8 to 11 deal with four of the 'second rank' of traditional hardware vendors. ICL is primarily a regional player, strong in the

UK and a number of other countries and in the vanguard of the open systems movement. UNiSYS is also strong in this area, and competes worldwide with IBM and Digital, albeit from a lower installed base. Hewlett Packard (HP) is particularly strong in technical and scientific computing. In the business computing arena, HP relies strongly on third party vendors to support its CASE strategy, CASEdge. The French-based Bull, like ICL, is primarily a regional player, and has a distinctive approach to distributed computing.

We did not consider the newer breed of RISC-pioneering, graphics-oriented UNIX 'box shifters', technically advanced and committed to open systems though they may be. Sun, MIPS and the like have not announced any CASE strategy comparable to the 'traditional' vendors.

However, there are a number of other hardware vendors which could have been examined, but for which there was neither the room or the time. Texas Instruments is a significant hardware vendor in its own right, but the authors believe that its activities as a software vendor are of more relevance, hence its inclusion there. Amdahl has its own distinctive and well thought-out strategy - Huron - which is positioned to compete with AD/Cycle, and would have been the obvious choice for one more chapter. Also, Fujitsu, Hitachi, Siemens/ Nixdorf and others might perhaps have been reviewed on the grounds of global or regional significance. But we believe, on balance, that we have done justice to the major vendors.

Chapter 5

IBM: SAA and AD/Cycle

1 Introduction

IBM finally unveiled one of the worst kept secrets in the history of computing on the 19th September 1989, when it announced its long term software development strategy. What had been anticipated was 'The Repository' but IBM's strategy – application development cycle (AD/Cycle) – encompassed CASE and methods as well. AD/Cycle provides the framework, within a systems application architecture (SAA) environment, for automating and integrating the phases of the development life cycle. The late Russell Jones stated that 'AD/Cycle legitimises the CASE and equally important methodology markets'.[1]

AD/Cycle provides a possible half-way house to solve the dilemma of interfacing versus integration. At the heart of AD/Cycle lies the repository – a dictionary that many products can use to share information, storing all the critical information relating to the development life cycle. Figure 5.1 shows the product environment for AD/Cycle as it was announced.

2 Systems application architecture (SAA)

Before looking at AD/Cycle it is necessary to understand its architectural context, systems application architecture (SAA). IBM offer systems based on several different hardware architectures and control programs. To make movement between these systems easier,

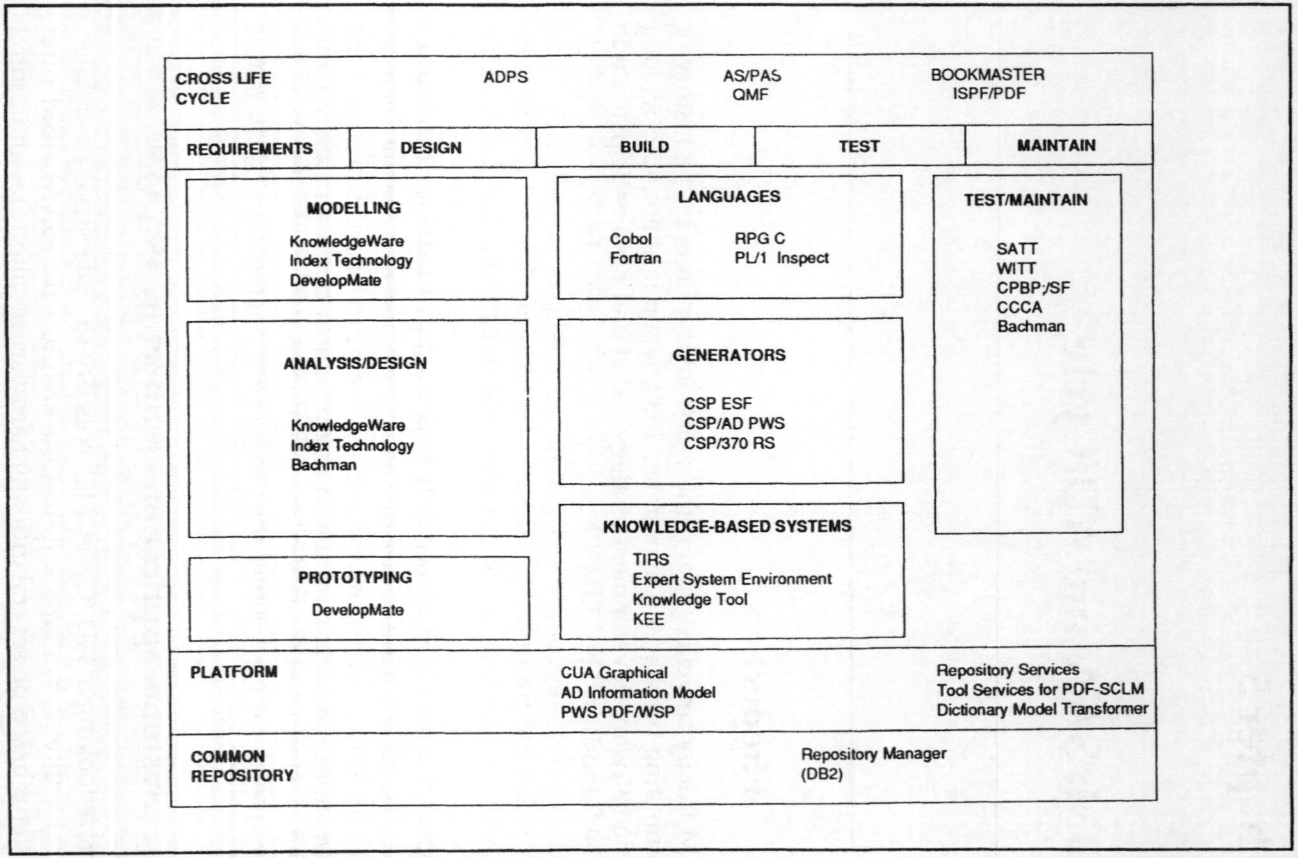

Figure 5.1 The AD/Cycle Product Environment

to facilitate the use of multi-user systems, and to bring the breadth of their product line to bear on user needs in all environments, IBM introduced SAA. The intended result is applications that can:

- be moved easily between systems
- utilise a simpler and more uniform user access.

SAA is a growing collection of selected software interfaces, conventions and protocols that are still being published at the time of writing, intended as the framework for developing consistent, integrated applications across the future offerings of the major IBM computing environments. The interfaces, conventions and protocols of SAA are designed to provide an enhanced level of consistency and connectivity in the following areas:

- programming interface – the languages and services that application developers use in building their software
- user access – the design and use of screen panels and user interaction techniques
- communications support – the connectivity of systems and programs
- applications – the software built and supplied by IBM and other vendors.

Figure 5.2 shows these components of SAA, and, schematically, how specialised and common solutions coexist.

Several combinations of IBM hardware and software have been selected as SAA environments: those in which IBM will manage the conformance to specification and the availability of support for all the SAA elements:

- TSO/E in MVS/ESA
- CMS in VM
- OS/400
- OS/2 extended edition
- CICS in MVS/ESA
- IMS/DC in MVS/ESA.

IBM state that SAA will be implemented fully in the TSO/E, CMS, OS/400 and OS/2 environments, whilst implementation in CICS and IMS/DC will focus on their host roles in host-with-independent-workstation applications, offering primary programming support and access to data.

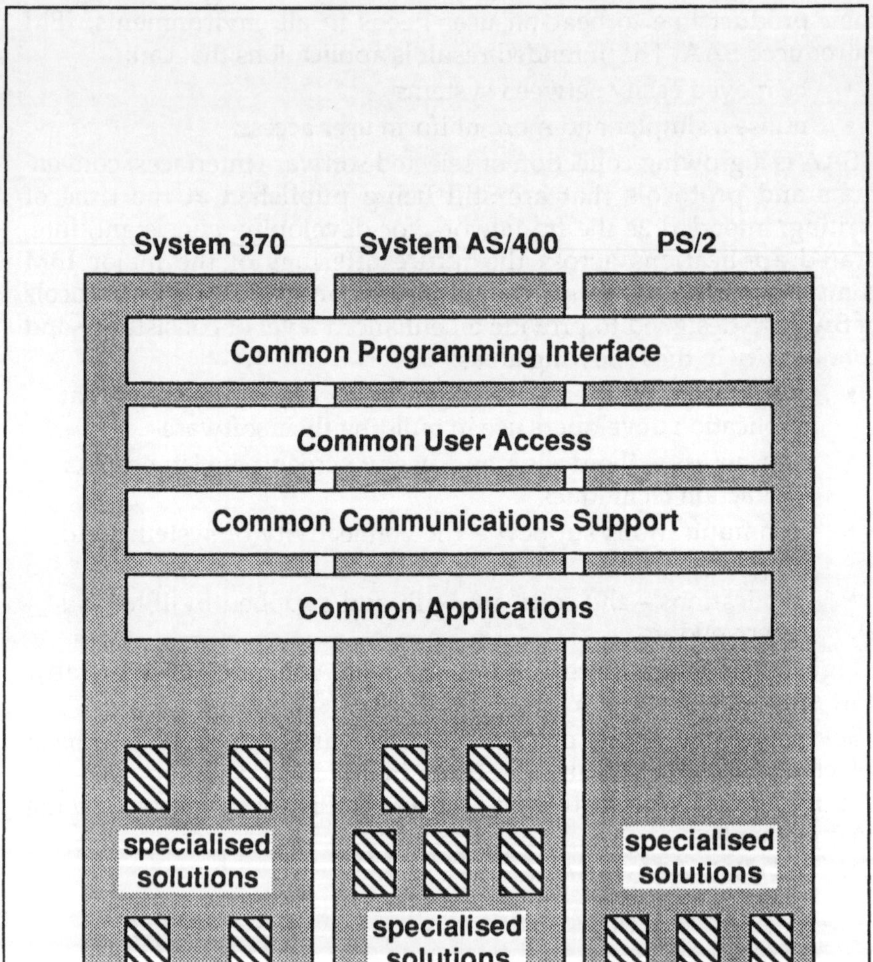

Figure 5.2 Systems Application Architecture (SAA)

2.1 Common user access (CUA)

The common user access (CUA) defines the rules for the dialogue between human and computer, establishing how information appears on a display screen, and how people respond to that information. It includes definitions of interface elements, rules for moving from one panel to another, choice selection, colour and emphasis, messages, help and terminology. The CUA consists of three main components:

- the way the machine communicates with the user
- the way the user communicates with the machine
- what the user understands about the interface.

The CUA interface deals with the following categories of consistency:

- physical – referring to the hardware: the keyboard layout, the location of keys and the use of the mouse
- syntactic – referring to the sequence and order of appearance of elements on the display screen (the presentation language) and the sequence of keystrokes to request actions (the action language)
- semantic – referring to the meaning of the elements that make up the interface.

The CUA lays down rules for interface elements and interactions, and gives application designers a fair degree of flexibility. It defines a number of panel types, also acknowledging that designers will require further, application-specific, types.[2]

2.2 Common programmer interface (CPI)

The common programmer interface (CPI) consists of the languages, commands and calls that programmers can employ. The components of the interface fall into two general categories:

- languages – consisting of application generators, C, COBOL, FORTRAN, procedures language and RPG
- services – consisting of the communications interface, the database interface, the dialogue interface, the presentation interface, and the query interface.

For each component of the CPI, IBM is establishing a definition or specification. These cover many of the aspects of a programmer's interaction with selected IBM software. The components in the specifications offer a broad span of function intended to meet most IT needs. Each component of the interface is an IT aid in itself, and is a tool that can be used on its own to get work done. In addition, many of these components can also be used to create larger, more complex applications to accomplish sophisticated ends. Figure 5.3 shows the structure of the applications and supporting interfaces.

2.3 Common communication support (CCS)

Common communication support (CCS) is used to connect applications, systems, networks and devices. This is achieved by the consistent implementation of designated communication architectures in each SAA environment. These communication architectures are the building blocks for the distributed function in the CPI and IBM SAA applications. Most of them are based on systems network architecture (SNA) and on international standards. Included in the CCS are:

- data streams – including 3270 data streams, consisting of user-provided data and commands, as well as control information that governs the way data is handled and formatted by IBM displays and printers; there are two significant components:
  - document content architecture (DCA), defining the rules for specifying the form and the meaning of a text document, providing uniform interchange of textual information in the office environment
  - intelligent printer data stream (IPDS), the high-function data stream for use with point-of-sale addressable printers
- application services, including:
  - distributed data management (DDM), the protocol architecture that defines access to record-oriented files located on remote SAA systems
  - SNA distribution services (SNADS), providing a synchronous distribution capability in an SNA network
  - document interchange architecture (DIA), providing a set of protocols to define several common office functions performed cooperatively by IBM products

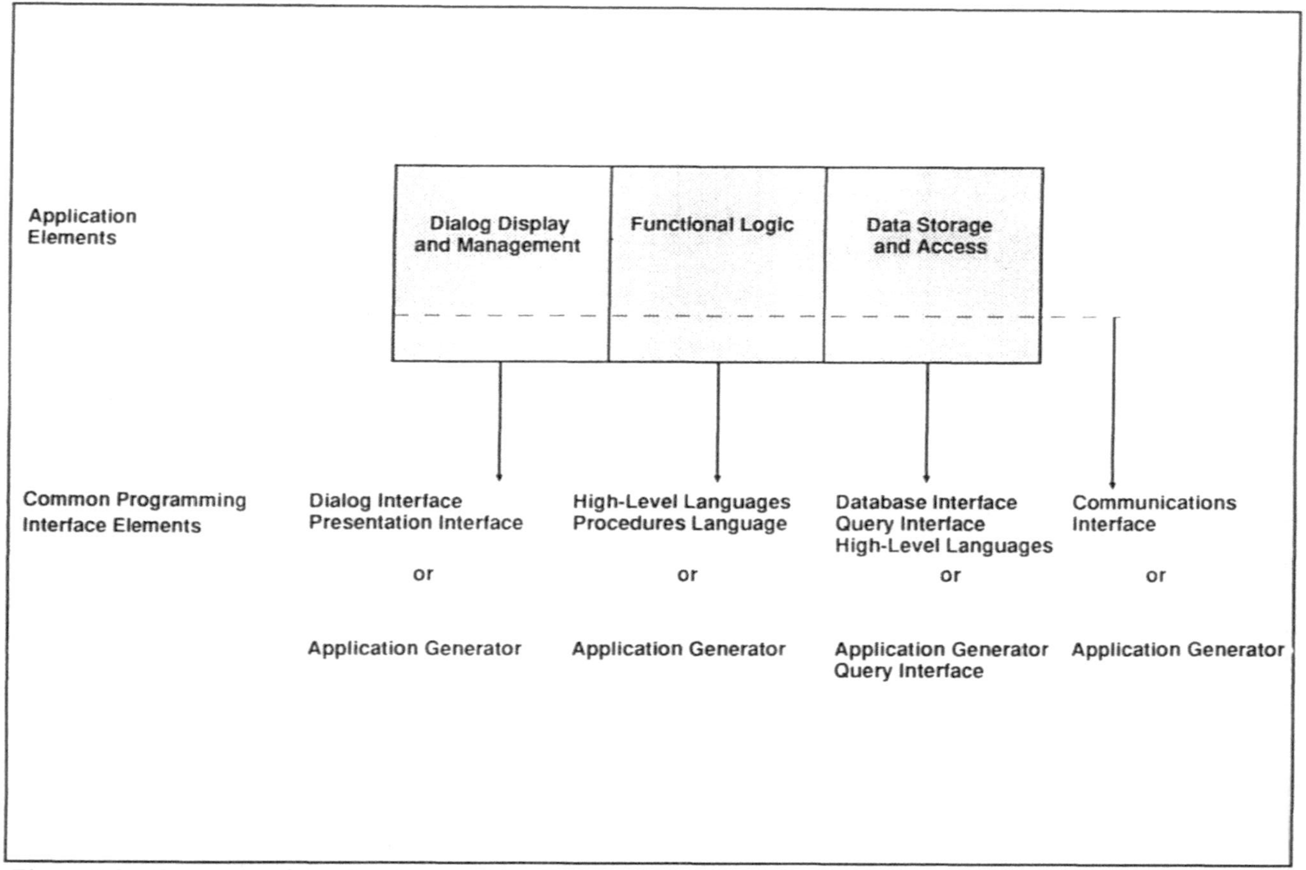

Figure 5.3 *Structure of Applications and Supporting Interfaces*

  - SNA network management architecture , describing IBM's approach to managing the network
- session services – LU type 6.2 is a program-to-program communication protocol
- network – low-entry networking nodes support peer to peer communication
- data link controls:
  - synchronous data link control (SDLC), a discipline for managing synchronous, code-transparent, serial-by-bit information transfer between nodes that are joined by telecommunication links
  - token ring networks, consisting of a wiring system, a set of communication adapters and an access protocol that controls the sharing of the physical medium by the stations attached to the LAN
  - X.25, defining the packet-mode interface for attaching data terminal equipment
- open system interconnection (OSI) protocols, covering the following:
  - FTAM, CCITT X.400 and ACSE (ISO 8650)
  - presentation layer (ISO 8623 and 8825)
  - session layer (ISO 8327)
  - transport layer (ISO 8073)
  - network layer (ISO 8473 and 8878).

2.4 Common applications

Common applications are applications developed by IBM using SAA. The first of these was OfficeVision, since discontinued, and the second AD/Cycle.

3 The content of the AD/Cycle announcement

The AD/Cycle announcement contained the following:

- a default life cycle method, IBSDM, shown in Figure 5.4
- a set of IBM software products centred around the repository and intended to support the default method

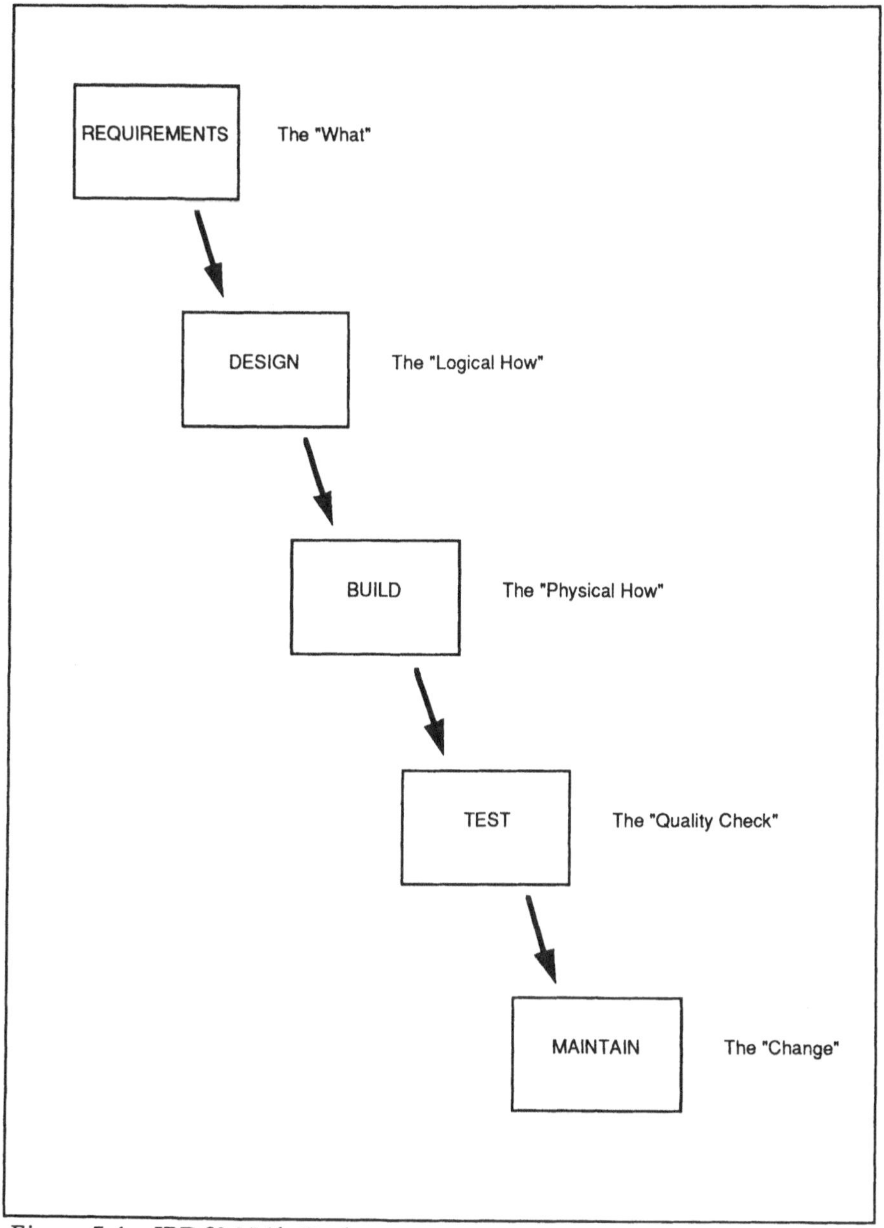

Figure 5.4 IBDSM Life Cycle

- a set of other third party software vendors whose products would be able to work with the repository and, in particular,
- a business equity link-up with 3 front-end CASE vendors:
  - Index Technology (now Intersolv)
  - Knowledgeware
  - Bachman.

IBM advise their clients to adopt a method which is well-defined, well-structured, technology independent and suited to business requirements. They provide their own solution, integrated business systems development method (IBSDM). This covers the five phases shown in figure 5.4.

4 Key AD/Cycle components

There are five key components of AD/Cycle:

- process manager
- repository
- application development platform
- method specific tools
- method independent tools.

The process manager provides a rule-based tool to enforce the chosen method, the sequence of tasks, the techniques and tools to be used, and the checkpoints. IBM claim that this controls process quality through management of the project tasks. This facility is provided by the ADPS (application development project support) tool.

The repository provides a logical single point of control and access for information, and structures the use of tools – IBM, third party and the users' own. This, say IBM, will provide quality information through the integration of data. The facility is provided through the repository manager (RM/MVS) and the repository CPI.

The application development platform provides an SAA/CUA end user interface and co-operative processing environment, as well as services for library management, and for data manipulation and validation. This should improve developer productivity by integrating functions, and by exploiting workstation technology. The programming interface is provided by the SAA CPI.

The method specific tools provide support for users' chosen development methodologies, and facilitate links to the business model, providing the ability to quality-control the application design early in the life cycle.

The method independent tools provide high-level language (3GL), application generator/4GL and knowledge-based systems technology support. They enhance productivity through task automation and through assisting maintenance.

These tools will sit on a platform that integrates techniques and methods, as well as IBM and other tools. IBM stated that its overall objective was 'to help you develop the right application, the right way for your business'. AD/Cycle exploits SAA, and therefore reaffirms that the latter remains a major strategic direction for IBM; in particular, AD/Cycle makes use of the SAA CUA interface.

The AD/Cycle announcement not only included IBM's own tools but those of third party vendors, who were divided into two classes: equity partners and participants. The initial equity partners were Index Technology (of Excelerator fame, and subsequently merged with Sage as Intersolv), Knowledgeware (of IEW and ADW fame) and Bachman (the reverse engineering CASE tool pioneers) and IBM took an approximate 5% equity stake in each. At the time of the announcement there were 32 participants. These included such well known names as CA, LBMS, Texas Instruments/JMA, Information Builders, Arthur Andersen and Softlab. The distinction between the classes is that IBM have set up marketing and technology arrangements with the equity partners. The participants will have access to AD/Cycle specification – in particular, to the tools interface and the information metamodel. During 1990, Synon and Systematica were added to the list of business partners. It is worth noting, in passing, that the business partnership scheme has not always guaranteed success for the companies involved: Systematica went into receivership in January 1992.

This linking of tools provides a half-way house between the choice of current I-CASE technology, which forces the user to work with only one vendor's products in a closed environment, and the 'pick 'n' mix' of interfacing CASE tools. Figure 5.5 compares the various approaches.

Current ICASE	**IBM Repository and satellites**	**Pick 'n' Mix**
(Tightly coupled)		*(Loosely coupled)*
'Thick' backplane:	**'Thin' backplane:**	**File interfaces:**
common model	common submodel of meta model	limited overlaps of metamodels
no data re-entry	data re-entry	semantic re-interpretation
potential for full cross stage services	limited potential for cross stage services	mostly one-way
potential for auto-transformations		
cross stage integrity	limited cross stage integrity	no cross stage integrity

Figure 5.5 AD/Cycle Compared to ICASE and 'Pick 'n' Mix'

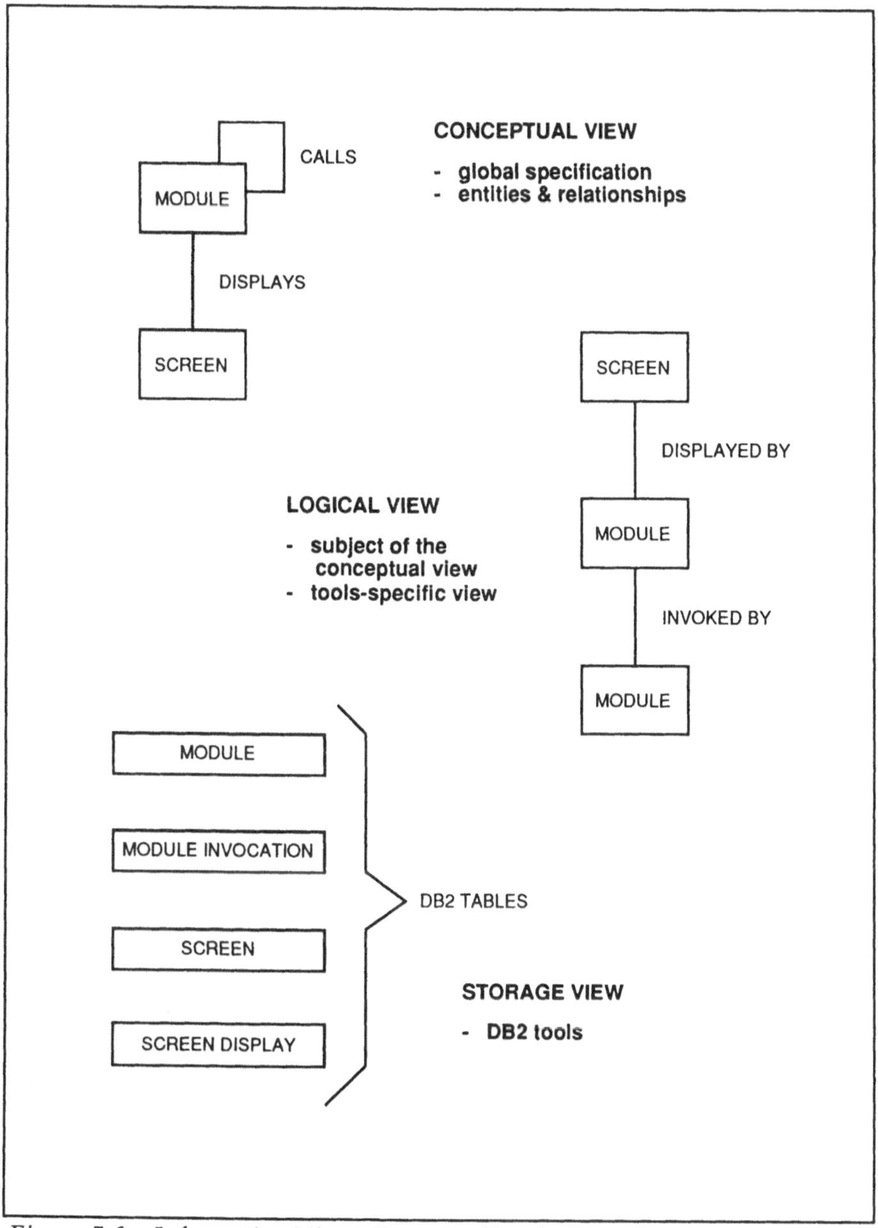

Figure 5.6 Information Views Supported by the Repository Manager

5 AD/Cycle products

The AD/Cycle products provided by IBM are as shown in figure 5.1. In this subsection we will look at the following major IBM products:

- the repository manager – RM/MVS
- ADPS
- SCLM
- DevelopMate
- SATT
- WITT.

5.1 The repository manager (RM/MVS)

The repository manager will use an SQL database engine as its storage manager. IBM stated that in its first release this engine would be DB2. However, the authors believe that this need not be the case, and that the repository can sit on any 'DB2 plug compatible' database software.

Repository users will not have to learn a new language to access its contents. IBM have extended QMF for this purpose, reducing the skill set needed to develop applications.

Figure 5.6 shows the information views to be supported by the repository manager.

5.2 Application development project support (ADPS)

ADPS originated in Germany as long ago as 1980, as a mainframe product called VIDOC. It was launched throughout Europe in late 1987 and first released in the USA in September 1989. It has had little uptake in the UK. Figure 5.7 depicts the ADPS architecture.

ADPS addresses the area of work management, which includes both project management and process management activities. This covers the definition of the activities to be performed, the work objects to be manipulated, and the conditions for performing activities. The current version does not conform to SAA CUA guidelines and at present does not have an interface to the repository manager (although this is likely to happen in the future). It has two components:

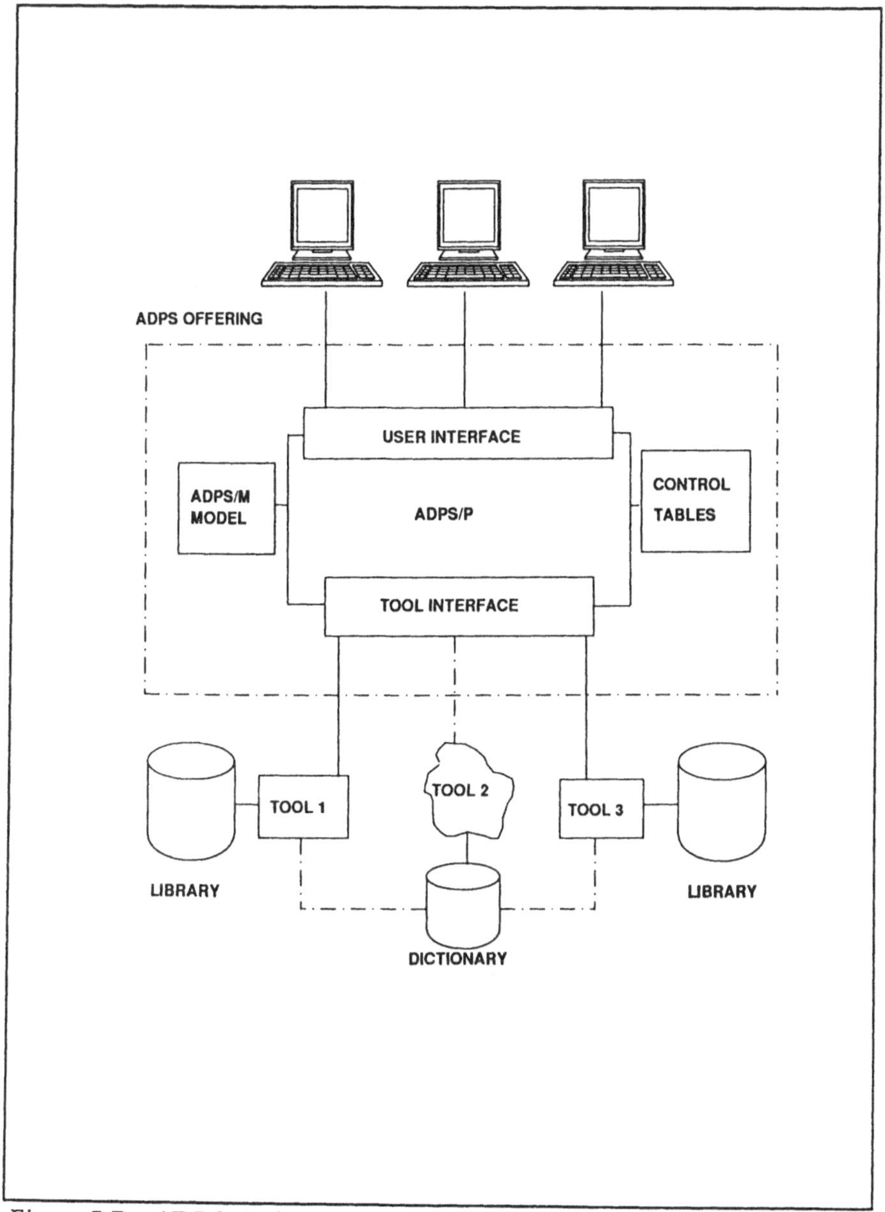

Figure 5.7 ADPS Architecture

- ADPS/M (application development process model) – a descriptive representation of the software development process, allowing an organisation to describe the activities to be carried out, the methods and tools to be used and the standards and guidelines to be followed
- ADPS/P (application development process mechanism) – the driver for the development process, containing dialogues which offer task selection for the developer through activities specified in the process model.

Figure 5.8 shows the elaborate five-piece host-workstation connection released in August 1990.

5.3 Source configuration library management (SCLM)

The SCLM product has facilities to aid the synchronising and management of source and object code. It can also be used to manage online screens, message scripts and so forth. Its build function allows the user to construct any sub-component of a system, or indeed an entire application. A project hierarchy can be described, which refers to the status of objects in the application development life cycle. SCLM is a mainframe base tool and figure 5.9 represents IBM's view of the interaction between SCLM and the workstation.

5.4 DevelopMate

DevelopMate provides the following:

- definition and maintenance of process and data modelling information in entity-relationship terms
- storage of modelling information in RM/MVS according to business model constructs
- interactive prototyping
- non-procedural specification of rules, sequenced during prototyping
- use of SAA interfaces and services
- German and English language support.

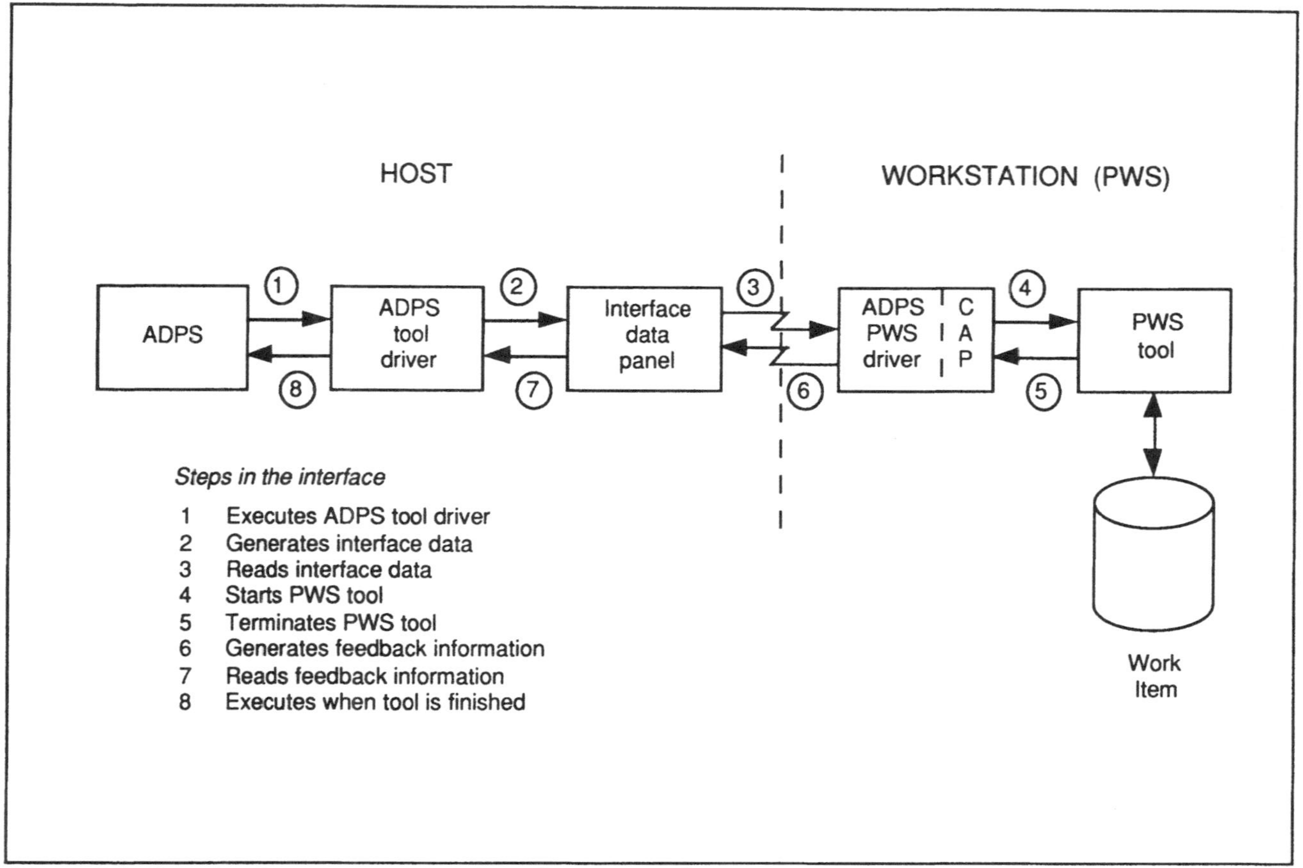

Figure 5.8 Interface Between ADPS and Workstation Tools

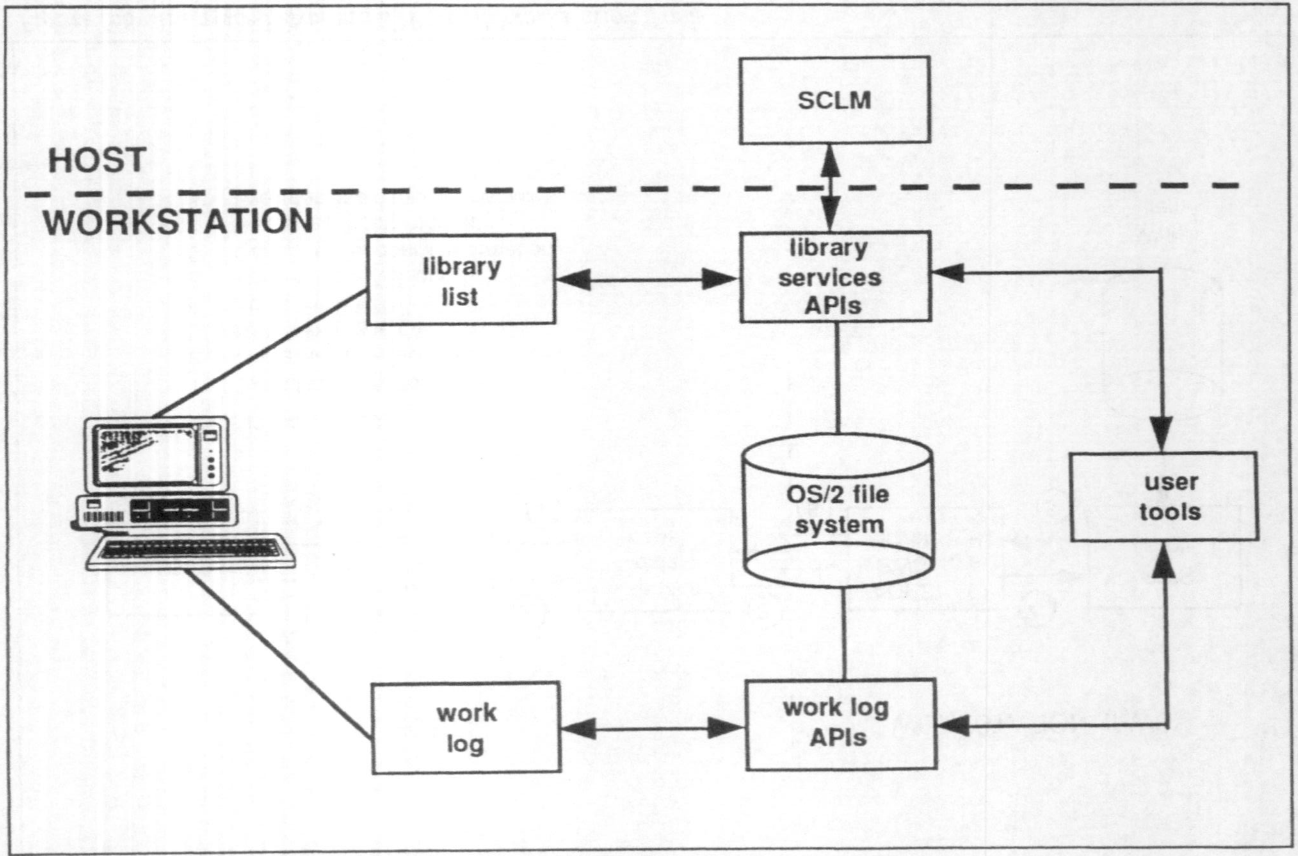

Figure 5.9 SCLM and the Workstation Platform for OS/2

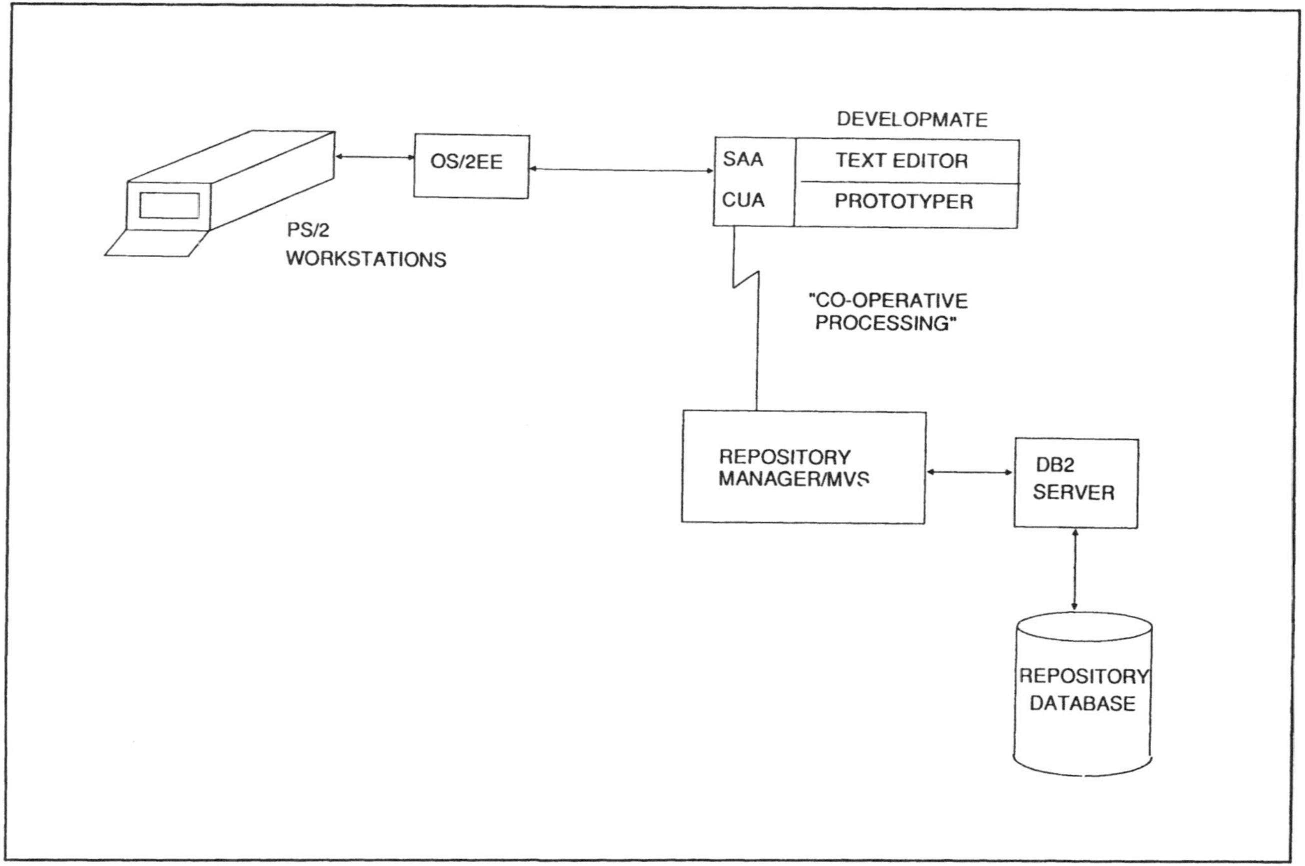

Figure 5.10 DevelopMate Environment

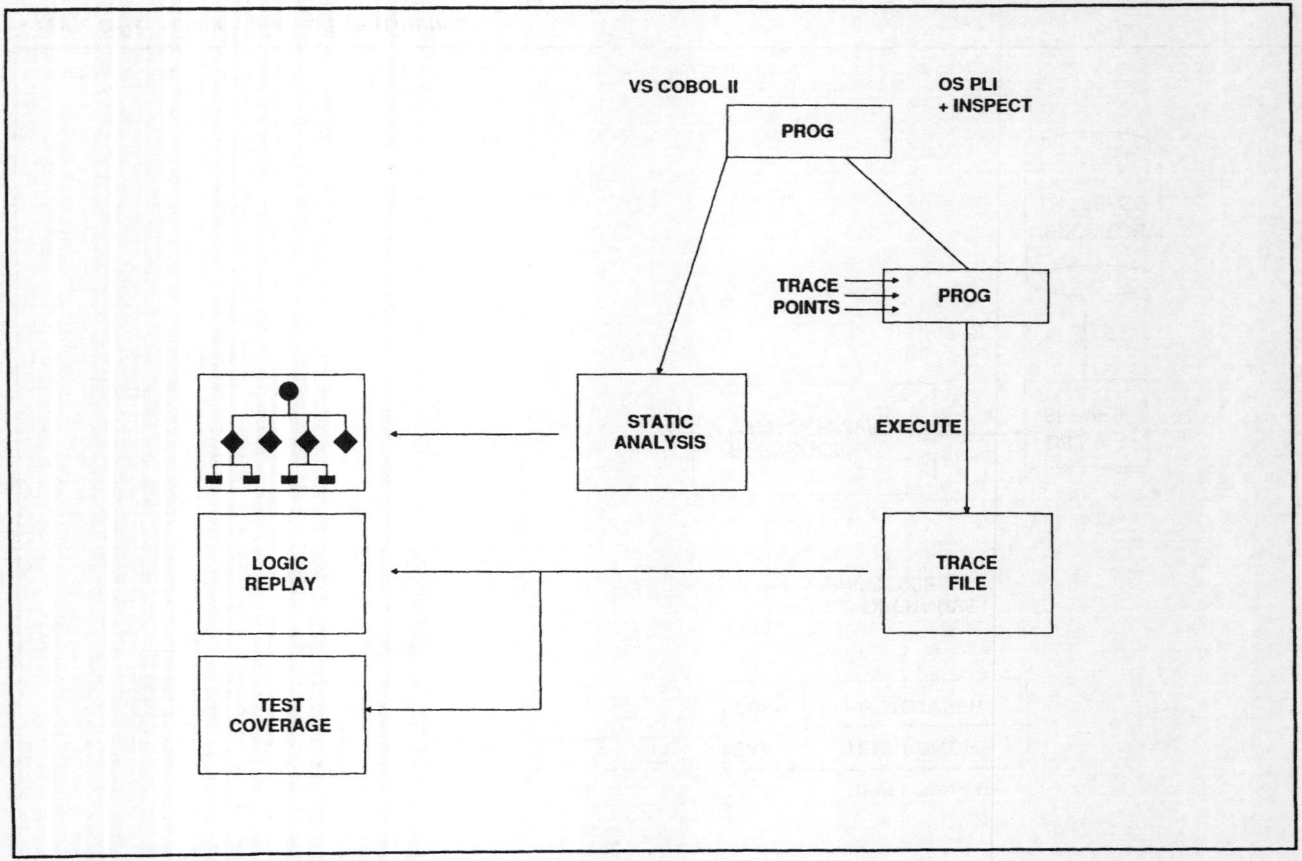

Figure 5.11 SATT Environment

DevelopMate is designed for use by business analysts, and allows a business model of processes and data to be defined in entity relationship terms. Business rules associated with attributes can be recorded. Prototyping specifications are defined at the workstation, whist execution takes place at the mainframe. DevelopMate provides no diagram support at the moment – a fatal limitation. See figure 5.10, which illustrates the DevelopMate environment.

5.5 Software analysis test tool (SATT)

SATT supports the developer in the testing and maintenance of applications. Working with IBM's language debugging facilities, SATT provides test coverage measurement and analysis support for the MVS/XA and MVS/ESA environments. It supports the following languages and subsystems:

- VS COBOL II, in:
  - TSO-E
  - CICS/MVS or CICS/ESA
  - IMS/VS or IMS/ESA TM
- OS PL/1 in:
  - TSO-E
  - IMS/VS or IMS/ESA TM.

SATT has an SAA CUA interface that supports browsing, querying and reporting of test coverage information, and has an animated test execution display facility. Figure 5.11 depicts the SATT environment.

5.6 Workstation interactive test tool (WITT)

WITT provides a regression testing function for applications in the following SAA environments:

- MVS/XA and ESA for TSO-e, CICS/MVS and IMS TM
- VM/SP and XA
- OS/400 TM
- OS/2 Extended Edition.

WITT provides facilities to automatically record and re-execute interactive application test sessions. An SAA CUA interface enables the query and display of test results and the analysis of test case discrepancies. An OS/2 Presentation Manager application provides host

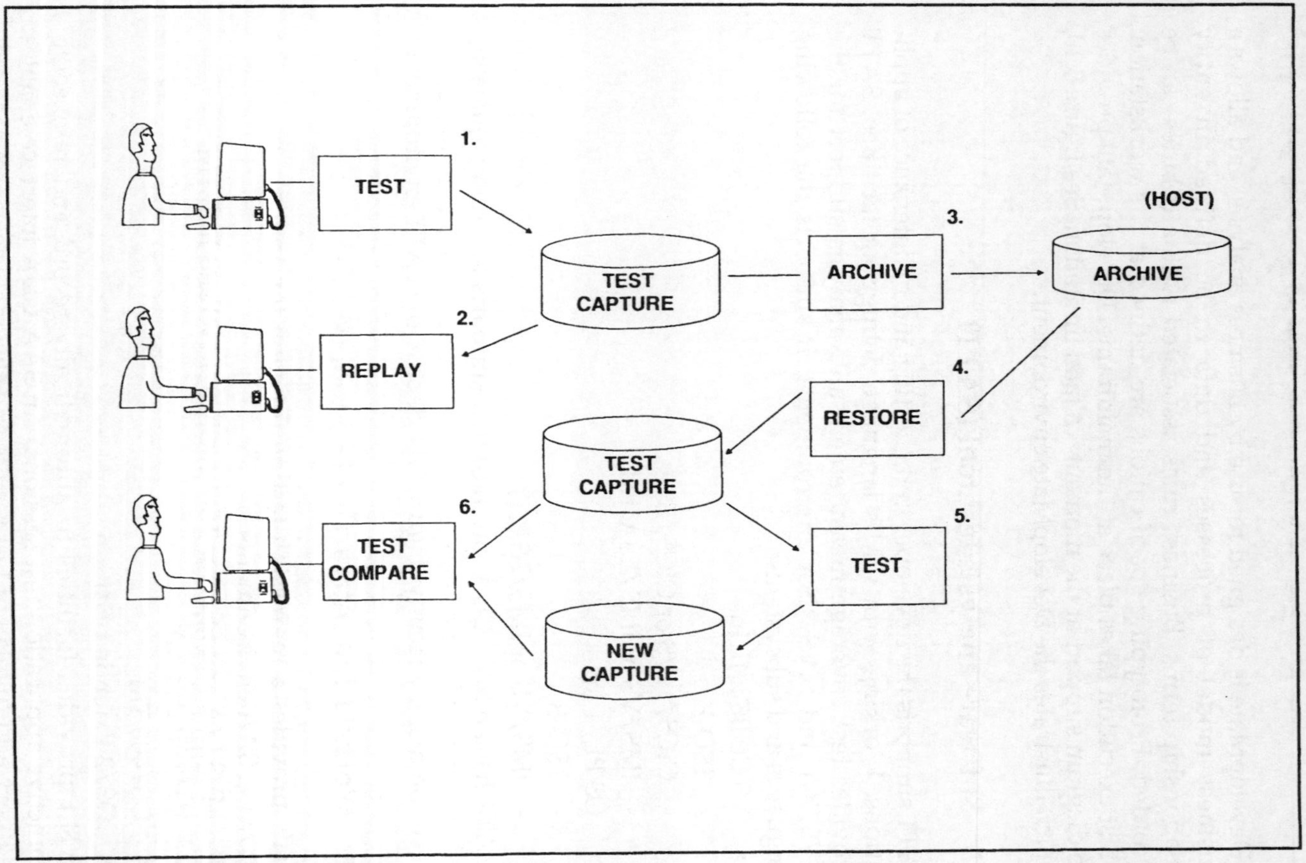

Figure 5.12 WITT Environment

and OS/2 application regression testing. Terminal sessions are saved in IBM programmable workstation disk files which can be sent to the host for storage using a built-in file transfer function. Figure 5.12 shows the WITT environment.

6 Issues not tackled by AD/Cycle

It is only by digging into the announcement, and starting to look at the implications, that the shortcomings of AD/Cycle become apparent. In the authors' view, three major issues are not addressed, namely:

- configuration management
- security
- management of live applications.

6.1 Configuration management

There are no configuration management facilities, not even basic support for versions. Configuration management can be defined as the process of formally establishing the manner in which all system elements (hardware, software or textual) will be maintained, and the procedures to be adopted when changing the state of any element. What is available from IBM?

- ISPF/PDF, through its SCLM feature, provides certain limited source code change control facilities
- two of the 32 participants (Legent and K3) provide configuration management facilities for source code on IBM mainframes.

However:

- DB2 has no support for version numbers
- IBM announced no link with its INFO/MAN product.

Configuration management is the key facility which can enable users to generate applications automatically, and to control the systems development life cycle; it is required not only for source and object code but also, without exception, for all other life cycle deliverables.

6.2 Security

The security of the information stored in the repository is more complex (if it exists) than needed. IBM state in the PLETS[3] that 'user management is responsible for evaluation, selection and implementation of security features, administrative procedures and appropriate controls in application systems and communication facilities'. The information to be stored in the Repository requires the same level of procedures and controls that would be given to business information. Therefore, IS/IT needs to analyse the risks to this information and manage those risks. The latter requires countermeasures. An examination of the AD/Cycle product set produces the following information:

- DB2 provides integrity constraints, permission, recovery, logging, data locking (at page level only) and views
- the repository manager (RM/MVS) is to provide facilities at the 'global level' and the 'tool specific level'
- RACF provides user identification, program/device permission, and password protection.

On the face of it there are facilities, but which should be used in what circumstances? Can the details be recorded – and coordinated – in one place? Unfortunately, IBM provides no guidance at all. The matter is further complicated by the absence of information on maintaining security with third party tools. In the authors' view, there is a strong likelihood that data integrity will be difficult to achieve if, indeed, it is at all possible.

6.3 Management of live applications

The repository is only active during the development process. Once an application is made live it passes out of the active control of the repository. Live application definitions will have to be passed to the DB2 directory and the CSP control file. This will duplicate, if not triplicate, the volume and complexity of meta-information – the scenario that information resource managers – such as data and database administrators – have been trying to avoid for the last ten years.

7 SystemView

In September 1990, IBM announced SystemView, its systems management strategy for planning, coordinating and operating heterogeneous, enterprise-wide information systems. This strategy addresses the management of hosts, workstations, personal system databases, storage, networks and business administration functions across the enterprise. It is the 'third dimension' of SAA. Figure 5.13 shows the three dimensions of SystemView.

As with AD/Cycle, IBM has provided a structure for the integration of its own and third party partner products. The SystemView partners are:

- Bachman Information Systems
- Candle Corporation
- Goal Systems International Inc
- Platinum Technology Inc
- Information Retrieval Company.

The SystemView philosophy is that systems management solutions should be consistent and easy to use, that they should share information and work together and that, in using them, IS/IT organisations can reduce costs through lowered training requirements and new automation opportunities. The three dimensions prescribe the way in which SystemView applications can interact and integrate:

- the end-use dimension provides an easy-to-learn consistent interface to SystemView applications
- the data dimension enables the sharing of information between applications, whether supplied by IBM or by SystemView partners
- the application dimension contains system management applications, providing the functionality to perform system management activities grouped into six focused categories – the SystemView disciplines; these disciplines, shown in Figure 5.14, represent the tasks performed in system management; they do not define new processes themselves, but rather facilitate processes an organisation may have already developed.

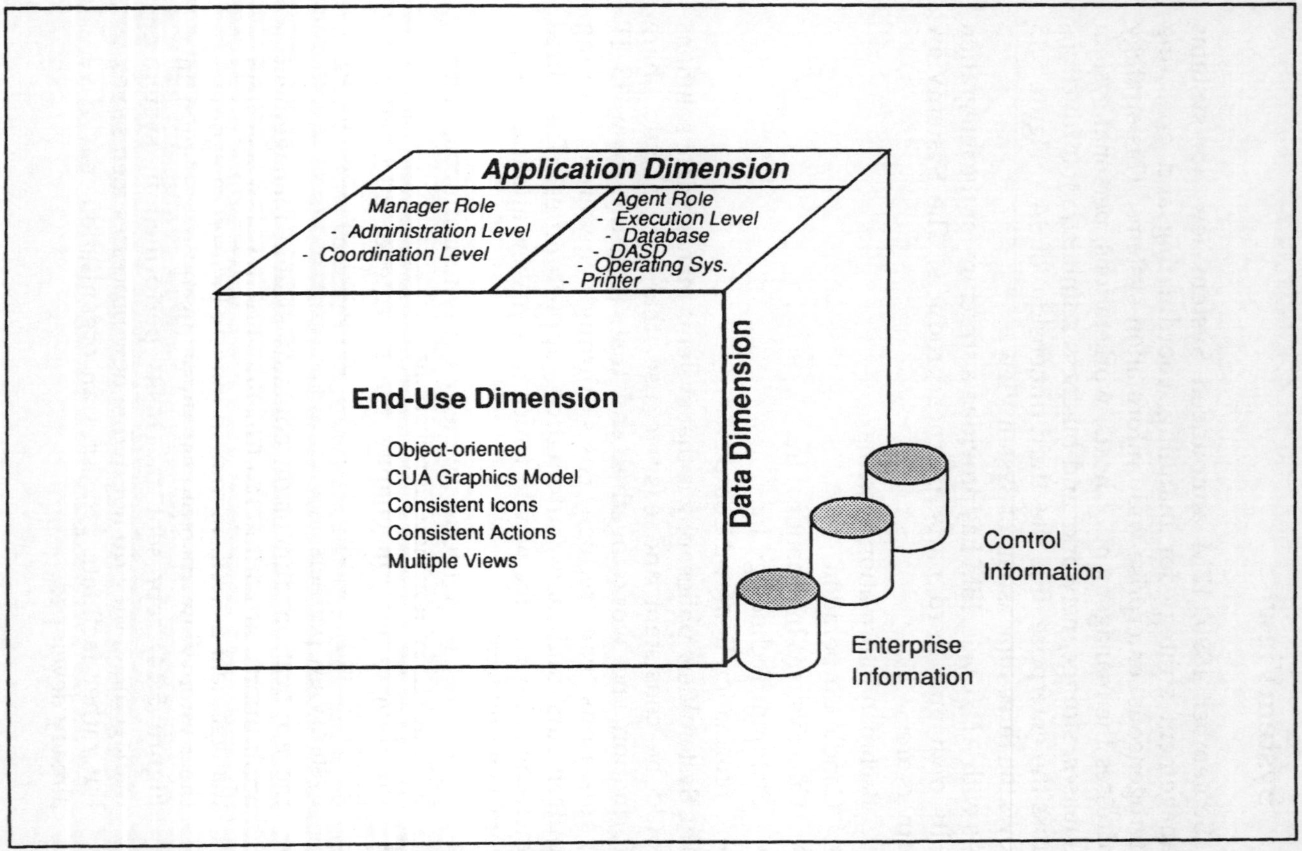

Figure 5.13 The Three Dimensions of SystemView

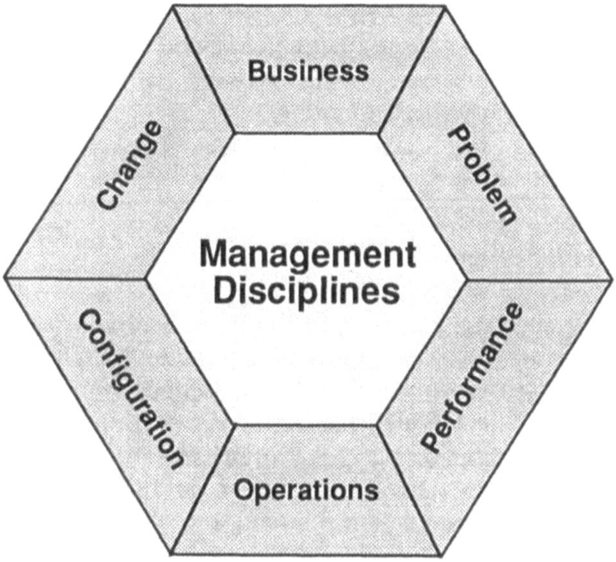

Figure 5.14 The Application Dimension Disciplines of SystemView

SystemView will need to be examined in conjunction with AD/Cycle, not least because it affects the management of the networks used in the development environment.

8 Information Warehouse

In September 1991, IBM announced the fourth SAA derivative, the Information Warehouse framework. This defines a set of database management systems, interfaces, tools and facilities that manage and deliver business information to authorised business users. The products in the Information Warehouse framework provide access to data wherever it is located in the enterprise – on IBM systems or on systems from other vendors. IBM has defined a framework of tools and facilities to manage and deliver 'timely, accurate and understandable' business information.

The Information Warehouse framework, as shown in Figure 5.15, is composed of three major parts:

- the enterprise data element
- the data delivery element
- applications and decision support systems.

The **enterprise data element** is the heart of the warehouse, and consists of a database management system which provides data integrity, security, recovery, reliability, availability and performance. Enterprise data includes all data that supports business operations and business decisions throughout an organisation, regardless of how that data is stored or where it is located. IBM's own DBMS products are designated for storing and managing data in the Information Warehouse, identified as follows for the various platforms:

- MVS – DB2 and IMS/ESA
- VM – SQL/DS
- VSE – SQL/DS and DL/1
- OS/400 – OS/400 Database Manager
- OS/2 – OS/2 Extended Edition Database Manager.

The **data delivery element** is the means through which the right business data is delivered to knowledge workers, and itself has a number of components:

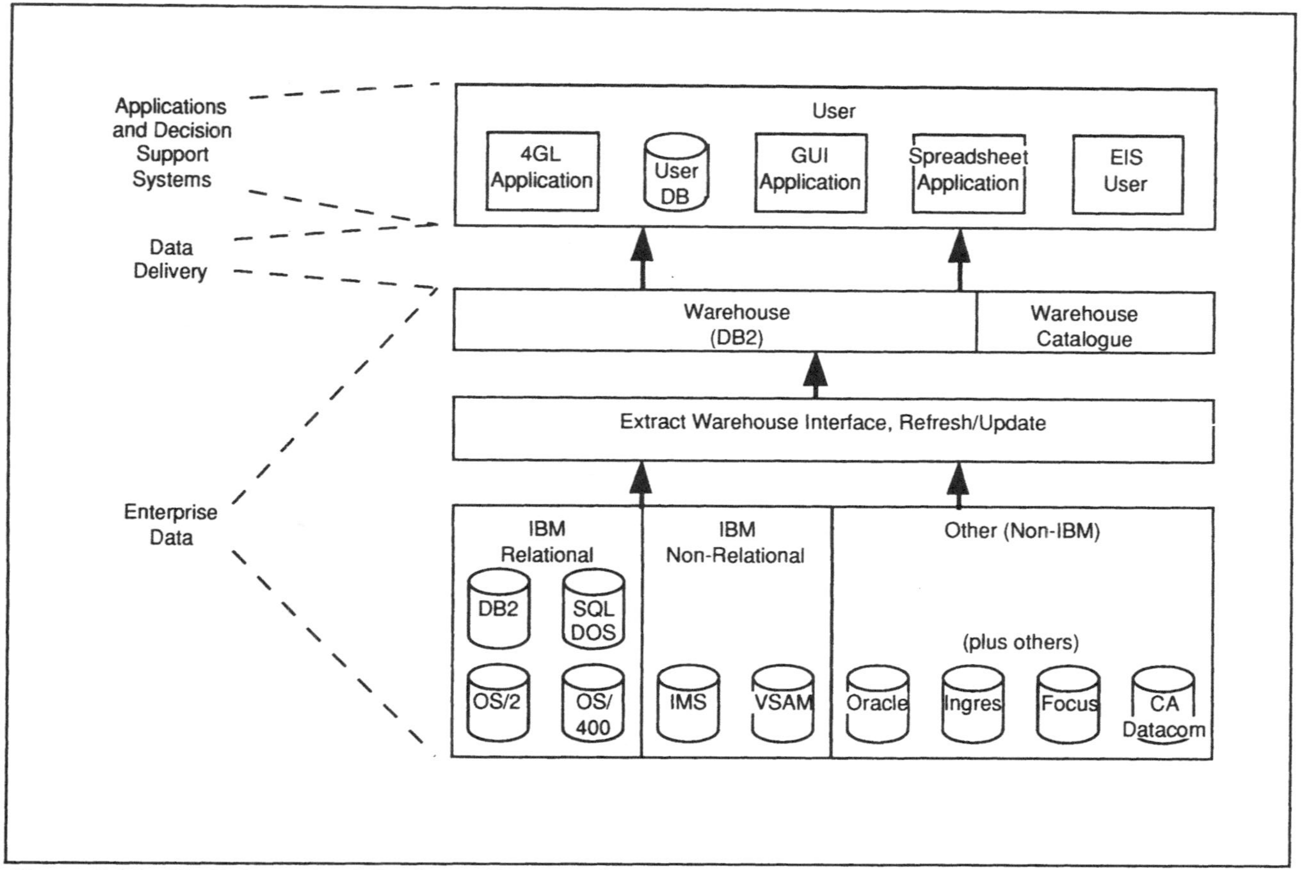

Figure 5.15 The Information Warehouse Framework

- data access – wherever that data resides in the enterprise
- data transport – extracting and propagating data from source files and placing it in other files for use by programs and end users
- data transfer
- data transformation
- data enrichment
- copy management.

Data delivery provides support for connectivity across various types of databases and files on different vendor platforms, and an essential part of this is a distributed database capability, provided by remote-unit-of-work support for DB2, SQL/DS and OS/400 Database Manager.

Decision support applications enable knowledge workers to transform data into information, by formulating a request for the selection of data, manipulating the data retrieved, and analysing and presenting the transformed data in various information formats and graphical charts. The Information Warehouse is designed to allow enterprises to unlock data which is currently difficult to access or unavailable to the knowledge worker. Products in this area relate to:

- query formulation and extraction to various databases
- natural language queries
- project management
- spreadsheets
- construction and transformation of graphical and image data
- EDI transfer of information
- executive support systems.

To implement the Information Warehouse concept, IBM has created another set of alliances with third party vendors; the first two were Bachman Information Systems, selected for its reverse engineering and data modelling experience, and Information Builders, which (in addition to FOCUS) markets a set of enterprise data access tools based on ESQL.

The Information Warehouse is a direct challenge to Digital's Information Network, which allows users to access IBM's DB2 database.

The framework is intended as a solution to business users' information needs. However, it may also be viewed as a solution for application developers, by viewing the repository as the database for the Information Warehouse. We feel that IBM is unlikely to adopt this solution to one of the main AD/Cycle deficiencies, and it will be up to consultancies and third party vendors to realise the potential.

9 AD/Cycle conclusions

IBM is asking its users, and potential users, to invest a considerable amount of money in the AD/Cycle product set. It is not unreasonable for users to expect some sort of return on that investment. Many complain that IBM has played the word game before, by promising that some piece of hardware or software is the universal panacea for all business and/or IS/IT ills. It must be admitted, of course, that IBM is not the only vendor to have done this.

Major questions surround AD/Cycle's ability to deliver results to system users. DB2 performance has been poor in the past, though with every new release IBM has improved the engine's efficiency. However, ADABAS, DATACOM/DB and MODEL/204 have consistently been able to outperform DB2 throughout its history, and downsizing, coupled with the reality of client/server technology, has brought Sybase and others into the picture as direct competitors.

One technical hurdle – data locking – has been ignored so far by IBM. Currently, DB2 locks data at the page level. A page, in IBM's terms, is 4K in size, and a fair number of records can fit on a 4K page. But it gets worse: not only is the data page size 4K but so is the index page size. In the distributed development environment of the 1990s the repository will probably be the most used database in the organisation, and there will be a high probability of two or more users wanting to access the same data. With the current locking capabilities, we foresee a physical limit to the number of concurrent repository users of between five and 20.

IBM identified a need for the products in AD/Cycle to be controlled in order to support the appropriate tasks in the development life cycle. This is nothing new, and has for a decade been the basis for the development of integrated project support environments – IPSEs. For this vital rôle, IBM has promoted the use of ADPS (see 5.2, above). The AD/Cycle environment, of course, supports not only IBM tools but also those from third parties. Unfortunately ADPS is a mainframe product, and only works with IBM products. There appears to be an impasse.

IBM's default life cycle is very old fashioned and does not reflect modern thinking. It fails to distinguish between a system development life cycle and a methodology, does not address the need for the development tasks to be related to the business strategy, and fails to comprehend that the concept of 'maintenance' is better addressed as performance tuning, error correction, enhancement, and so forth.

We question how customisable the information metamodel and the supporting object services are. This is vital, because the support provided for organisations' life cycles and methodologies is dependent on the answer. Unfortunately, it was in this very area that the AD/Cycle announcement gave little information, and nothing since has served to enlighten us.

Within AD/Cycle, the repository promises users a means of combining the advantages of an ICASE tool's single dictionary, with the open architecture benefits of a 'pick 'n' mix' approach. To achieve this desired state, it is important to reach agreement on the contents of the model, and also on the services that need to be provided. These big questions, in our view, have not been dealt with by IBM, neither in the initial (1989) announcement nor at any subsequent stage.

Finally, IBM may be outflanked by ... IBM! In January 1992, the company announced its industry-standard SDE (software development environment) strategy for the RS/6000 UNIX world, as described in Chapter 6. Paul Hessinger, chief technology officer of Softlab and an AD/Cycle adviser, was reported as suggesting that SDE may yet eclipse AD/Cycle: 'IBM does not yet realise how much SDA will force AD/Cycle to live or die'.[4]

10 References

1 Perfectsure Ltd (date uncertain), *Software Development Monitor*

2 More detailed information on the CUA can be found in IBM Manual SC26-4351

3 IBM AD/Cycle PLETS: ZA90-0130, ZP90-0148, ZP90-0146, ZA89-0236, ZP89-0375, ZP89-0376, ZP89-0384, ZP89-0386, ZP89-0385, ZP89-0380, ZP89-0378, ZP89-0387, ZP89-0388, ZP89-0379, ZP89-0237, ZP89-0382,ZP89-0374, ZP89-0373, ZP89-0372, ZP89-3061

4 Computing (1992), *IBM unveils details of case strategy for RS600 range,* (article), 30th January, 1992
5 IBM (1989), *Application Development Announcement Summary,* 19th September 1989
6 IBM (undated), *Partners in AD/Cycle,* G320-9872-00
7 IBM (undated), *Getting started – planning for IBM's AD/Cycle,* G320-6707-00
8 IBM (undated), Applications Development Project Support: A better way of working, G511-0455-0
9 IBM (undated), *AD/Cycle; Blueprint for a more productive future,* G520-G-626-00
10 IBM (undated), *DevelopMate General Information V.1.R1,* GC26-4641-0
11 IBM (undated), *Software Analysis Test Tool General Information V.1 R.1,* GC26-4671-1
12 Advanced Software Development Research (1989), *Zap Analysis,* September 19th 1989
13 Perfectsure Ltd. (1989), *Software Development Monitor,* September 1989
14 Perfectsure Ltd. (1989), *Software Development Monitor,* October 1989
15 IBM (undated), *AD/Cycle Roadshow Presentation Script*
16 Holloway, Simon (1990), *Souped-up Cycle,* IBM Computer Today, May 23rd 1990
17 Yourdon, Ed (1990),*The Second Coming of AD/Cycle, American Programmer,* vol.3, no.11, November 1990

Chapter 6

IBM: AIX CASE

1 Introduction

IBM announced advanced interactive executive (AIX) CASE shortly after they publicised their RS/6000 RISC-based machines and a new version of the UNIX operating system, AIX version 3. It consists of IBM and a number of business partner products which support software development. Their message is that AIX CASE addresses the twin objectives of increasing a developer's productivity while supporting software engineering methodologies, distributed workstation development environments and open systems. This is achieved through five key technological components:

- advanced compiler technology – extended C, Fortran and Pascal compilers
- software development life cycle tools – supporting CASE tools for specific phases and across the life cycle
- advanced user environment – using the X-Windows interface and the OSF/MOTIF graphical user interface
- object oriented technology – supporting C++ and using object oriented technology for the visual interface and for tool integration
- a tool integration framework – defining specifications and conventions.

Figure 6.1 uses an AD/Cycle picture to indicate the software supported by AIX CASE.

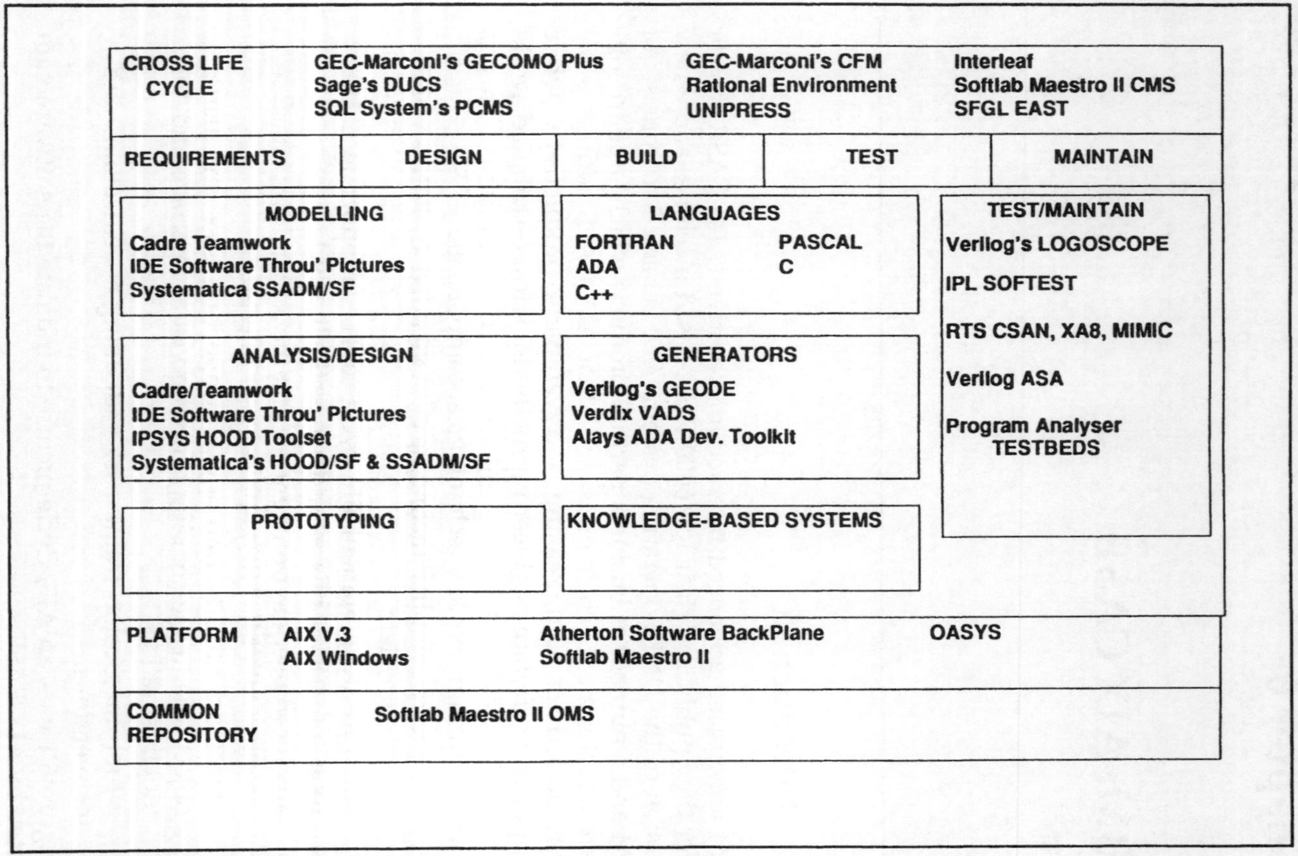

Figure 6.1 AIX CASE Structure

2 AIX CASE solutions

IBM has pre-packaged a number of CASE solutions for AIX. These are based upon language requirements, project size and complexity, and the need for an integrated environment. Each of the AIX CASE solutions allows users to integrate their existing tools, which can be added using the following additional components:

- Interleaf technical publishing software
- Token Ring and Ethernet LANs
- Atherton Technologies' offerings – Software Backplane and IPSE .

Various AIX CASE solutions are offered:

- foundation group
- FORTRAN
- C
- C++
- embedded C
- integrated C
- integrated ADA.

The foundation group is the base level entering point, and consists of the AIX version 3 operating system, the base application development toolkit – BADT/6000 – and the AIX windows environment/6000. The toolkit comprises:

- source code control system (SCCS)
- enhanced symbolic debugger (dbx)
- X development environment (xde)
- Assembler
- standard UNIX tools.

The FORTRAN solution provides a set of products for FORTRAN development, consisting of the foundation group, the AIX XL FORTRAN compiler/6000 and a run time environment. In addition, users are offered the following complementary software :

- AIX computer graphics interface (CGI) toolkit
- optimisation subroutine library (OSL).

The C solution provides a development environment for C language developers. It consists of the foundation group, the AIX XL C compiler/6000, and Saber Software Saber-C.

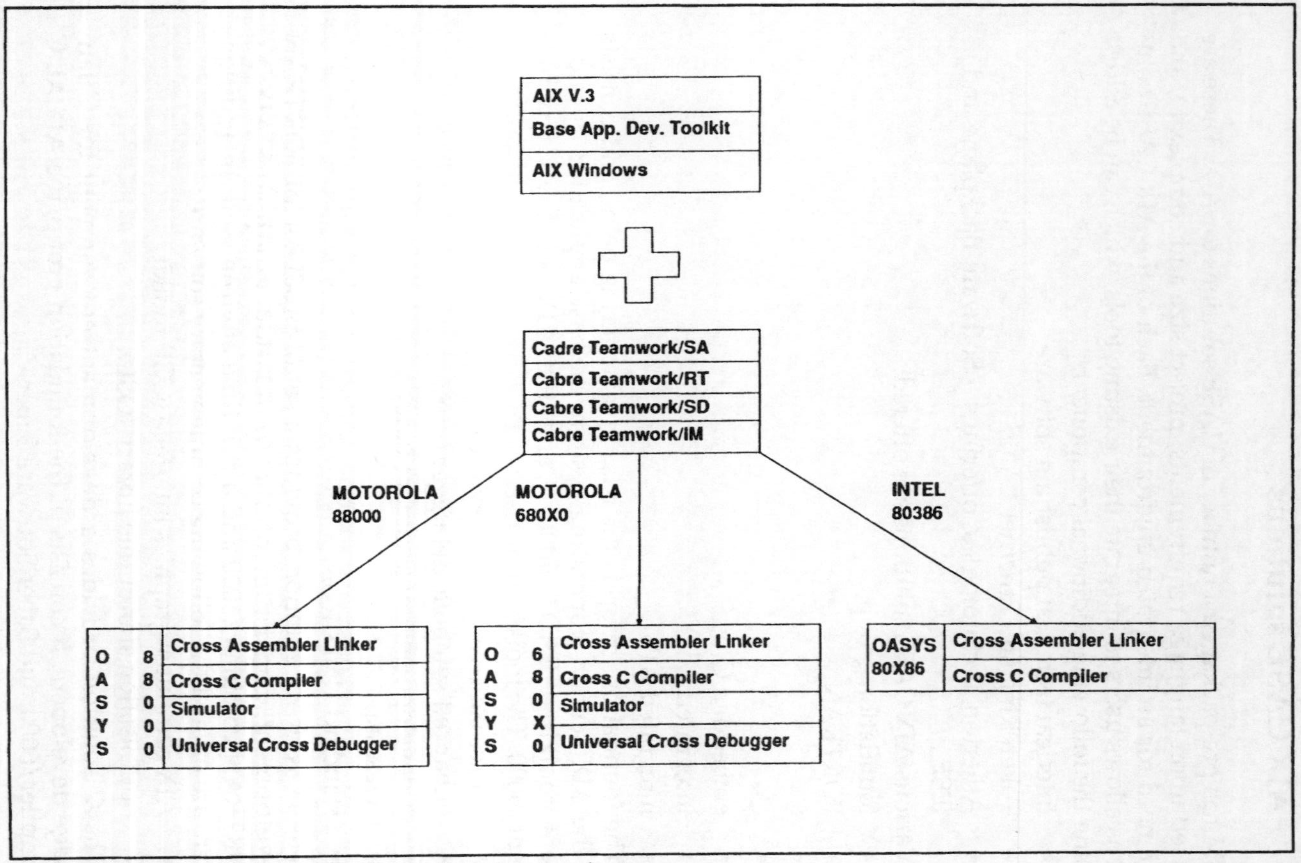

Figure 6.2 AIX CASE Embedded C Solution Products

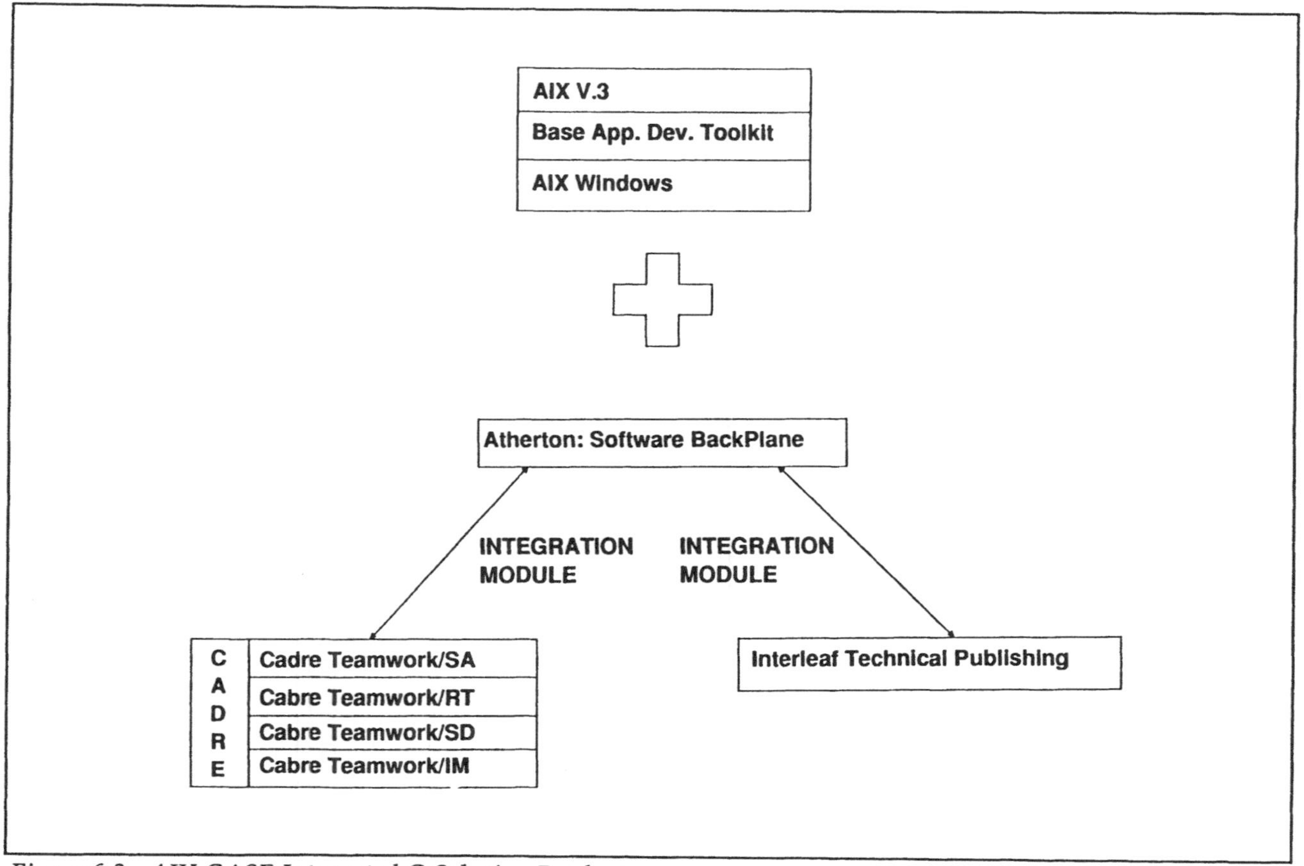

Figure 6.3 AIX CASE Integrated C Solution Products

The C++ solution provides an object oriented development environment for the C++ user. It consists of the foundation group, AIX XL C compiler/6000, OASYS' Glockenspiel C++ and Screens++, and the Unipress C++ MACS editor.

The embedded C solution provides cross development for a choice of microprocessors. Figure 6.2 shows the products offered in this solution.

The integrated C solution provides an integrated, structured environment for large scale C language development projects. Figure 6.3 shows the products offered in this solution.

The integrated ADA solution provides an integrated, structured environment for large scale ADA language development projects. Two alternative ADA development and compilation environments are offered: IBM's own or Verdix's VADS. Figure 6.4 shows the products offered as part of this solution.

3 Cross life cycle and life cycle support

Cross life cycle tools provide functions across multiple stages of the development life cycle. These activities include those that apply control or extend access to information across multiple users. AIX CASE provides the following areas of support:

- process management – controlling the process of application development by defining a standard set of procedures, identifying the sequence of tasks, the tools to be used and appropriate checkpoints
- project management – producing critical path, Gantt and pert charts, and performing time and resource checking
- configuration management – helping to organise, construct and maintain software
- documentation management – covering the control of documents to support life cycle deliverables
- editors – providing contextual help, aiding debugging, providing text windows and full access to UNIX utilities.

Cross life cycle support is provided by a number of IPSE products such as Rational, Atherton Technologies' Software Backplane and Softlab's Maestro II. Configuration management tools supported include SQL Systems' PCMS and GEC-Marconi's CFM.

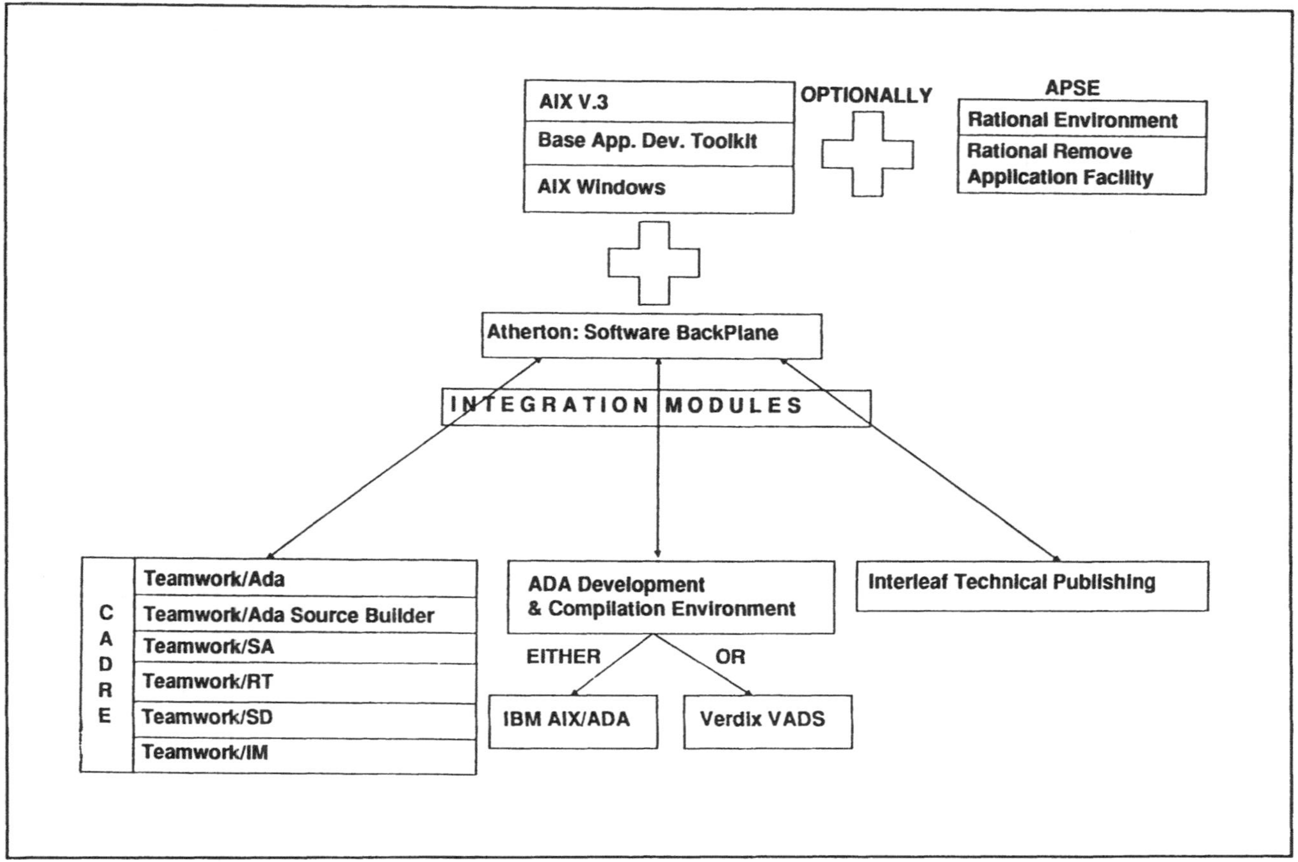

Figure 6.4 *AIX CASE Integrated ADA Solution Products*

Life cycle tools address particular stages in the development life cycle. In AIX CASE, these stages are :

- analysis and design
- language support
- testing and maintenance.

AIX CASE uses products such as Systematica's IBSDM derivative of VSF* and the Cadre Teamwork product set to support analysis and design. Language support is handled by the various language compilers available. Application testing is supported by products such as Program Analysis' TESTBEDS and IPL's SOFTEST.

4 Comparisons with AD/Cycle

As a vendor, IBM has strengths in almost all areas of CASE strategy – but prototyping is an exception, unfortunately. The curious fact is that to benefit from certain of the strengths you need AD/Cycle, while for others AIX CASE is the only option. Very few areas are strong in both strategies, and IBM has yet to address this paradox. AIX CASE is strong in the weak areas of AD/Cycle and weak in some of the strong areas. Figure 6.5 uses the AD/Cycle structure diagram components as a means of comparing the two strategies.

5 AIX CASE conclusion

AIX CASE is an open environment which takes advantage of UNIX cross life cycle support tools. The strategy is based on open, international standards. In Atherton Technologies' Software Backplane there is a potential industry standard, as Digital has also adopted the product. AIX CASE is currently weak in support of commercial IS/IT CASE tools, but IBM is encouraging third party vendors to join in. ORACLE has already ported its CASE product set to AIX version 3, and AIX CASE supports the concept of distributed development.

AIX CASE may not have the publicity behind it that has made AD/Cycle almost a household name, but IBM UNIX users should find it of more practical value than its better known 'rival'.

---

* Note that Systematica went into liquidation early in 1992.

AD/CYCLE COMPONENTS	AIX CASE	AD/CYCLE
REPOSITORY	THIRD PARTY PRODUCTS (MAESTRO II, RATIONAL)	MVS IBM PRODUCTS SYNON FOR AS/400 NOTHING FOR PS/2 GENERALLY UNSATISFACTORY
PLATFORMS	BASED ON INTERNATIONAL STANDARDS X WINDOWS, OSF/MOTIF, UNIX ATHERTON TECHNOLOGIES' SOFTWARE BACKPLANE	IBM PROPRIETARY SAA BASED, PARTICULARLY CUA REPOSITORY SERVICES STILL NOT CLEAR
CROSS LIFE CYCLE (PROCESS MANAGER)	STRONG THIRD PARTY SUPPORT FOR EDITOR DOCUMENTATION, PROJECT MANAGEMENT AND CONFIGURATION MANAGEMENT	ADPS AND AS FROM IBM A VERY WEAK AREA
MODELLING AND ANALYSIS / DESIGN	BUSINESS PARTNERS AND THIRD PARTIES MAINLY REAL TIME BASED EUROPEAN IPSE-BASED SUPPORT FOR COMMERCIAL IS/IT	IBM, EQUITY PARTNERS AND THIRD PARTIES COMMERCIAL IS/IT
PROTOTYPING	NO REAL SUPPORT	IBM'S DEVELOPMATE A WEAK AREA
LANGUAGES	FORTRAN, PASCAL, ADA, C AND C++ THIRD PARTY COMPILERS SUPPORTED	SAA COBOL, FORTRAN, RPG, C AND PL/1 IBM ONLY
GENERATORS	THIRD PARTIES	IBM ONLY
KNOWLEDGE BASED SYSTEMS	TIRS	IBM AND THIRD PARTY SUPPORT
TESTING AND MAINTENANCE	THIRD PARTY PRODUCTS NO REVERSE ENGINEERING	IBM AND LIMITED THIRD PARTY SUPPORT

Figure 6.5 AIX CASE Versus AD/Cycle

6 References

1 IBM (undated), *Advanced Interactive Executive /CASE*, sales publication G320-9939-0

2 IBM (undated), *AIX CASE*, PLET, ZP90-9004

Chapter 7

Digital: NAS and COHESION

1 Introduction

Digital Equipment Corporation (DEC) initiated its CASE environment strategy in 1988 with the release of an extensible dictionary, CDD/Plus. Index Technology (now Intersolv) was the first third party vendor to make use of CDD/Plus as an alternative information repository for Excelerator. This use of CDD/Plus by third party vendors was encouraged from the start by Digital, and was emphasised in September 1989 in its CASE environment announcement – one week after IBM's AD/Cycle saw the light of day. In June 1990, Digital further refined its approach to incorporate network applications support (NAS), and announced a strategy named COHESION. CDD/Plus was subsequently renamed CDD/Repository, to bring it into line with current terminology, and to emphasise that Digital has a repository *now*.

Digital's vision for software development is to enable organisations to develop and deploy software for multi-vendor environments, for all industries and markets, for all sizes of project and for all styles of computing. This vision is built upon NAS (for readers interested in knowing more about NAS, contact Digital). Figure 7.1 shows the CASE environment relationship to NAS.

Digital's CASE environment is comprised of the following elements:

- services and support, consisting of Digital consultancy to aid users in the selection and implementation of CASE tools

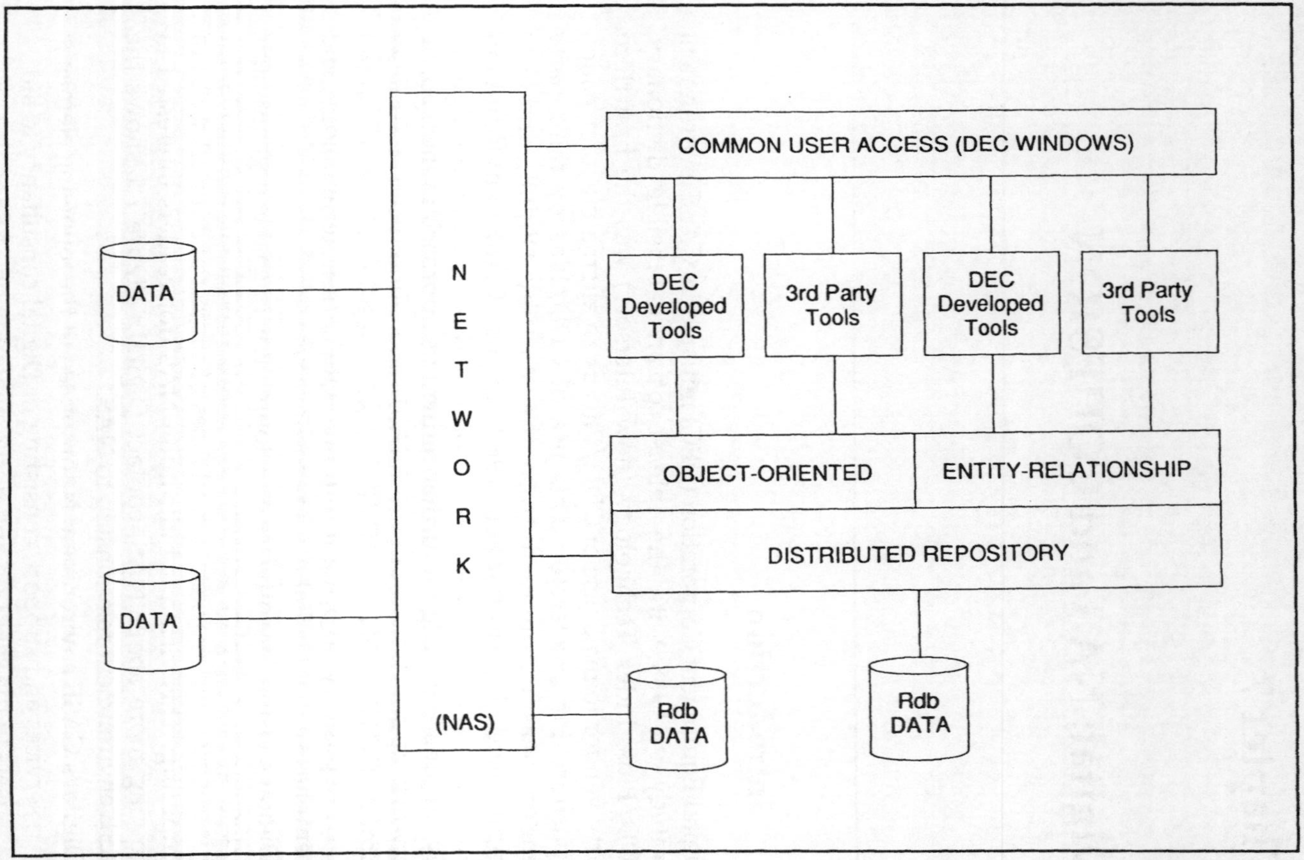

Figure 7.1 Digital's CASE Environment

- a set of architectural standards-based interfaces and components, based on NAS, to enable multi-vendor software development and the integration of applications
- a unified software development environment – the integration framework – designed to enable tool integration
- a family of tools, used to automate tasks within the development life cycle and across that life cycle.

2 Network Application Support (NAS)

NAS is central to Digital's vision of multi-vendor distributed computing. This vision is one in which application integration is accomplished across the full spectrum of computing requirements – transaction processing, database management, CASE, office systems, and computer integrated manufacturing (CIM). To achieve this level of integration, NAS delivers a set of application services that are:

- based on existing and emerging vendor-independent industry standards
- built in accordance with the NAS architecture for interoperability and portability of applications.

These services are built on top of Digital's database, networking and systems architectures, and serve to separate an application from these underlying concerns. NAS services consist of code that is delivered as developers' toolkits, runtime software libraries, and the like. Both Digital and independent vendors can utilise NAS services to build applications that access, communicate and share information and resources across a network of different desktop devices and server systems. The services are modular, and include:

- application access services, to provide users with transparent access to applications across a network
- communications and control services, to provide common facilities for applications to transport messages within an organisation and between organisations
- information/resource sharing services, to allow applications to share information and system-wide resources
- system access services, to provide a common operating system interface.

The architecture behind NAS has been 10 years in the making, and has evolved from the original single-system architecture approach in the VAX VMS machines. The key principles underlying NAS are externally focused, and include:

- open, multi-vendor operations
- distributed operation
- interoperability and portability
- flexibility and protection of investment.

2.1 Open, multi-vendor operations

NAS implements a number of open specifications for interfaces, services and supporting formats, to enable applications to be ported across a wide range of systems with minimal changes, to interoperate with other applications on local and remote systems, and to interact with users in a style that facilitates user portability.

The interfaces that comprise NAS are based on existing or evolving vendor-independent standards. This adherence to standards provides for a common understanding of the treatment of key components, including application programming interfaces, data representations and communications protocols. Figure 7.2 shows the variety of platforms that NAS can be used and implemented on. NAS services are committed to run on both the TCP/IP and DECnet networks, and in the future will run on DECnet/OSI.

2.2 Distributed operation

NAS components support distributed applications whose parts run on different nodes in a network; neither the nodes nor the networking protocols need be homogeneous.

One of the techniques used to achieve this is the client/server mode of distributing computing. This partitioning of services enables a formal application programming interface (API) to define all data and control-passing between the client and the server. The client/server distribution is handled transparently. Other communication services underlying NAS include Digital's distributed naming services, which provide a means to name various objects in the network in a manner that is independent of the location of the namer.

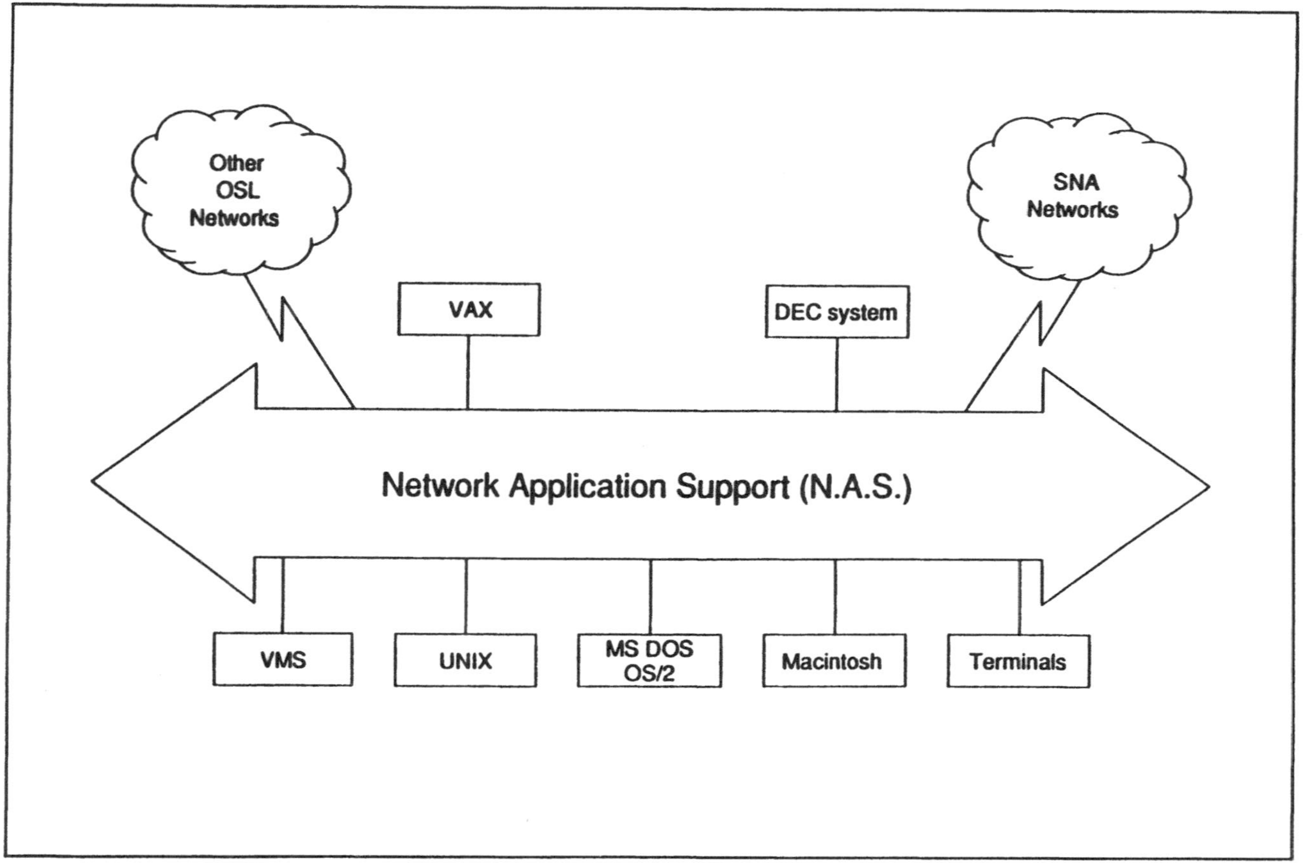

Figure 7.2 NAS Multi-Vendor Platform

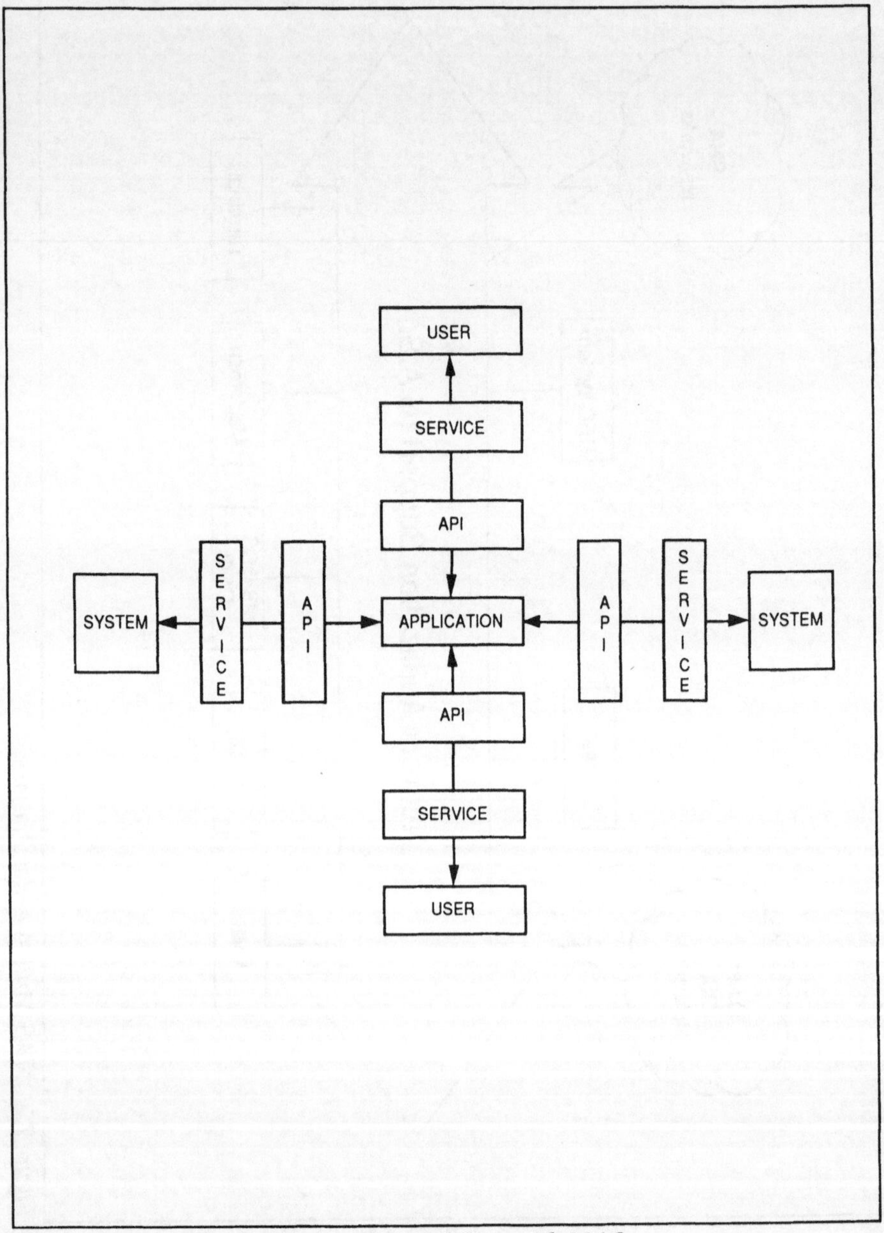

Figure 7.3 Interoperability and Portability with NAS

2.3 Interoperability and portability

Both interoperability and portability are facets of working in a unified software environment.

Interoperability refers to the ability to distribute the functions of an application across a network to different hardware/software platforms. Like portability, interoperability can be seen to function in levels. The degree to which computers from multiple vendors work together progresses from the hardware level to the application level. Interoperation at the highest (application) level requires, in addition, interoperation at the lower levels.

Portability refers to the ease with which an application can be moved from one hardware/software environment to another. The degree of portability varies with the degree to which applications are coded with system-specific or non-standard features. Digital's current emphasis is on portability between NAS server platforms.

Figure 7.3 illustrates the relationship between these two aspects of NAS. The common set of APIs and services that run across different platforms provide a level of platform transparency for an application.

2.4 Flexibility and protection of investment

The ability to accommodate change while protecting existing investment in hardware and software is a trademark of Digital's approach to computing. NAS helps insulate applications from changes in the underlying technology by providing a set of higher-level services. Lower levels can change, while the application interface remains stable.

2.5 Adherence to standards

All of the above depend on adhering to industry standards when developing applications for a multi-vendor environment. Portability and interoperability of applications across heterogeneous systems is possible only if the multiple platforms have a common ground or set of conventions. The NAS services follow standards of many types, and their adoption is prioritised as follows:

- *de jure* standards stipulated by accredited standards-developing bodies such as ISO, CCITT, X3, IEEE or ANSI

SERVICE	RELEVANT STANDARDS
APPLICATION ACCESS WINDOWING	X WINDOW SYSTEM (X/OPEN; ASC) OSF/MOTIF (OSF; IEEE 1201)
FORMS	FORM INTERFACE MANAGEMENT SYSTEM (ISO SC22/WG18)
TERMINAL GRAPHICS	ANSI X3.64 GKS AND PHIGS (DEC) ISO 7942 AND 9592 ANSI X3.124 AND X3.144
APPLICATION CONTROL	REMOTE PROCEDURE CALL (NCS/RPC; OSF RFT; ECMA 127; ISO DIS 10148; ASC X3T5) IEEE 1003.4 AND 1003.8
COMMUNICATION AND CONTROL MESSAGING	CCITT X.400
EDI	ASC X12 EDIFACT
INFORMATION/RESOURCE SHARING COMPOUND DOCUMENT	ODA/ODIF (ISO 8613) SGML (ANSI X3.143; ISO 8879) ABSTRACT SYNTAX NOTATION (ASN.1-ASC X3T2, ISO 8824-5)
DATA ACCESS	SQL (ISO; ANSI X3.135)
REPOSITORY	IRDS (ANSI X3.138; ISO TC97)
FILE SHARING	NETWORK FILE SYSTEM (NFS) (DE FACTO; IEEE 1003.8)
PRINT	ECMA/ISO DISTRIBUTED PRINT ARCHITECTURE POSTSCRIPT (DE FACTO)
SYSTEM ACCESS	POSIX ANSI/IEEE 1003.1; ISO 9945-1

Figure 7.4 Industry Standards Supported by NAS

- consortia specifications, such as those adopted by the Open Software Foundation and X Consortium
- *de facto* standards utilised in the industry such as PostScript and network file system (NFS)
- ad hoc standards, such as those represented by X and CASE Integration Services
- Digital corporate standards, such as the compound document architecture.

Figure 7.4 shows the specific industry standards incorporated into NAS services.

3 Architectural standards adopted in COHESION

The relationship between NAS and the CASE environment is bidirectional, for not only is the CASE environment built upon NAS, but also it is used to build NAS applications. Standards are used by NAS in order that developers can build applications which are portable, integrate well with other applications, and can be distributed across a network of computer hardware and software from different vendors. These standards include:

- X-Windows, the standard behind DECWindows (in addition to Digital's announced support for OSF/MOTIF through the same product)
- ATIS, a proposal to the ANSI and ISO IRDS standards bodies for tool integration, and the basis for the new CDD/Repository object-oriented interface
- SQL – the relational database standard from ANSI and ISO
- POSIX 1003.1 – this is the IEEE standard for operating system interfaces
- CCITT X.400 – message transport services
- CCITT X.23 – public network services
- TCP/IP – UNIX protocols for networks.

The Digital CASE Environment takes advantage of the integration services provided by NAS through its use of standards. This enables tools and applications to be integrated across an organisation.

4 The COHESION family of products

The tools in the COHESION Environment are divided into two major categories. The first set is addressed at automating the work of building and changing applications. This set are called the life cycle tools. The second set enables the management of the processes and procedures that deal with applications. These are called the management tools. Figure 7.5 shows the Digital products that fit into these categories.

4.1 Life cycle tools

CDD/Repository is an evolution of the CDD/Plus dictionary product, and provides a mechanism for integrating data management functions and life cycle tools on both VAX/VMS and RISC/ULTRIX platforms. The scope of the product has expanded from storing information about data (records, fields, elements and procedures), to storing information about objects (anything involved in software development or information management, such as modules of code, databases, models, files or documentation), and methods (operations that can be performed on one or more objects such as create, edit, display).

CDD/Administrator is the first of a series of repository administration and access tools. It will allow users who manage repositories to locate, query, manipulate, manage and report on objects stored in CDD/Repository through DECWindows (MOTIF-based) on VAX VMS and RISC ULTRIX platforms. It provides multiple styles of customisable navigators to quickly locate objects stored in CDD/Repository, and customisable editors to display, modify or create objects in CDD/Repository.

DECdesign is a front end CASE tool, derived from Systematica's VSF, that supports a multiple set of analysis and design techniques, including YSM (with Ward-Mellor extensions), Gane and Sarson, and MERISE, and automatic code generators. Users can extend and customise the product to meet their requirements. It runs under DECwindows on VAX VMS.

SSADM-SF and HOOD-SF are VSF method-specific, single user front end CASE products that run on VAX/VMS and RISC/ULTRIX.

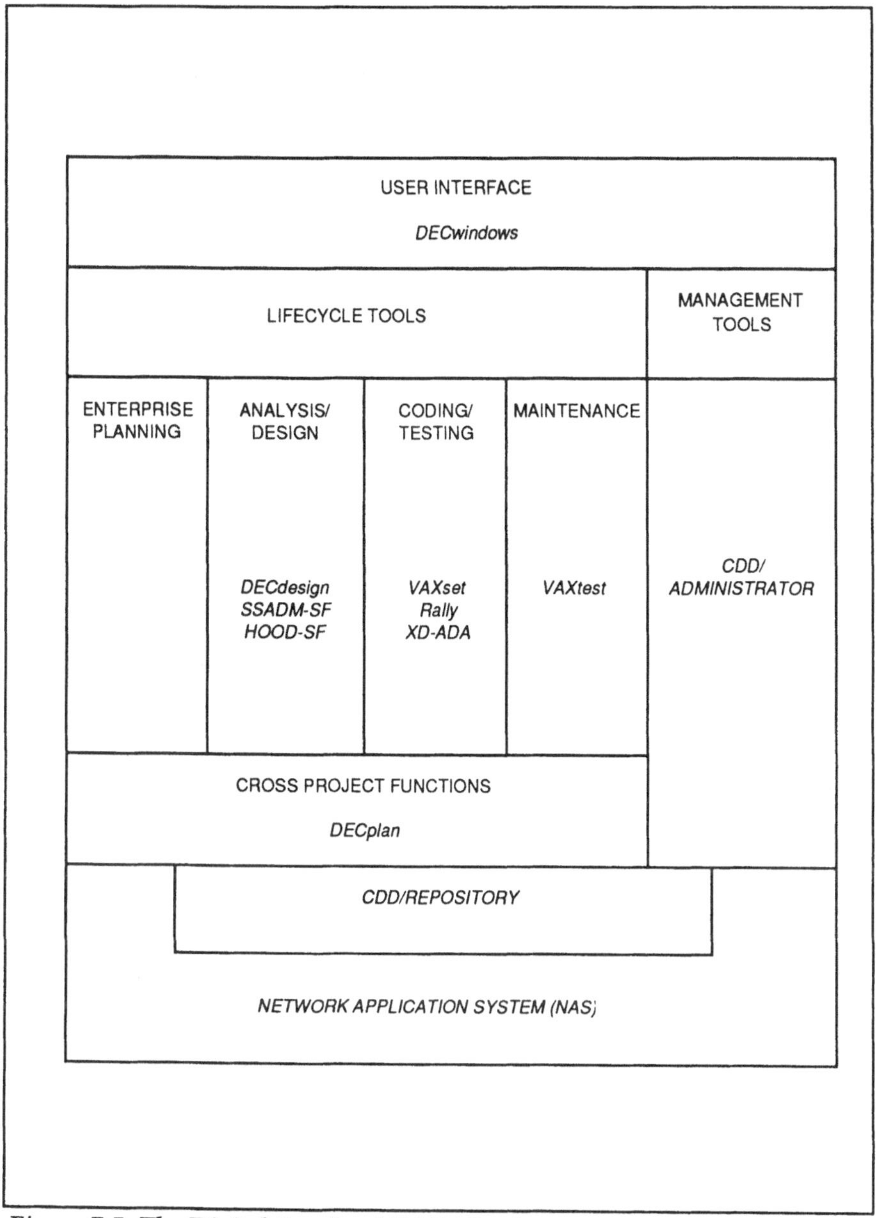

Figure 7.5 The Digital Family of CASE Products

4.2 Process management tools

DECplan merges the disciplines of time and project management in a workstation or DECwindows terminal environment to produce an activity management solution. It provides a common set of tools that allow individuals or workgroups to manage and control their time, resources, costs and deliverables. DECplan consists of several components, and is based on a client/server architecture. There are two servers – one for VMS and one or RISC ULTRIX – which can share data, and each client can connect to them. There are two clients per NAS platform – DECplan and DECplan Time Manager. The DECplan server performs all the calculations that they need, and stores the data. The DECplan client provides the user interface for the project and time management capabilities of the system, representing each of these capabilities by a set of views. The DECplan Time Manager client provides electronic calendar facilities.

DECset (originally called VAXset) is a package of software development tools. It is a programming environment that operates across both VAX VMS and RISC ULTRIX, and can be used to build and maintain all types of applications. It consists of:

- language sensitive editor (LSE), an editor with a built-in knowledge of VAX VMS languages (C, ADA,. Fortran, Pascal, COBOL, PL/1, Basic, SCAN, BLISS and DIBOL) and RISC ULTRIX languages (C, Fortran and ADA)
- source code analyzer (SCA), to perform cross-referencing and static analysis on design information
- performance and coverage analyzer (PCA), to pinpoint runtime bottlenecks and to determine sections of code not executed during testing
- test manager (TM), to automate the administration and running of program test procedures and the evaluation of test results
- code management system (CMS), providing an automated librarian and version management system
- module management system (MMS), a software system builder designed to automate the development of application systems
- program design facility (PDF), a set of enhancements to LSE and SCA to support module design, including pseudocode.

DEC FUSE (Friendly Unified Software Environment) is a UNIX programming environment for RISC ULTRIX workstations and servers. It is based on UNIX commands and utilities, and makes extensive use of dynamic graphical capabilities. DEC FUSE integrates these tools and provides a uniform DEC OSF/MOTIF-based user interface to all tools. It supports C, Fortran and Pascal program development, and consists of the following tools:

- editor, providing mouse/button-based editing
- debugger, a graphically enhanced UNIX debugging facility, capable of transforming a set of 'dbx' commands into push buttons
- program builder, based on the UNIX 'make' utility, to serve as an aid in compiling and building application software
- call graph browser, providing a mouse-sensitive graphical display of the call hierarchy of application software
- profiler, providing post-processing capabilities for RISC ULTRIX 'prof', 'pixie' and 'pixstats' utilities for program optimisation
- cross-referencer, providing a window-based, mouse-sensitive tool for locating and classifying named program elements
- code manager, providing source code control capabilities to support programming work groups.

Tool integration in DEC FUSE is accomplished with a multicast message server (MCMS), which maintains a list of message patterns that each tool can send and receive.

Rally is Digital's strategic 4th generation application development environment. The latest version (2.2) introduces the open data interface (ODI), which allows users to integrate Rally applications with 3GL code to access information from any source including VAX ACMS database servers and Rdb.

The VAX COBOL generator is a terminal-based, screen-oriented program generator that produces VAX COBOL source programs. It was originally shipped in 1986. Digital has entered into an agreement with Touch Technologies for the maintenance, support and future development of the product, which Digital will continue to sell.

XD-ADA, a family of ADA cross-compilers, is a joint development between Digital and SD-Scicon. It is based on the VAX ADA compiler front-end and SD-Scicon's code generation technology.

Digital Trellis is an object-oriented application development system, which includes a strongly-typed object-oriented language, an interactive tool kit, SQL database support, persistent workspaces and a type library.

Digital is in the process of defining a repository information model, taking a bottom-up and top-down 'two pronged' approach. CDD/Plus provides a model of bottom-up development, produced to support products using the dictionary. Digital is also working top-down on the future view of the information model. This effort will cover the complete life cycle including conceptual, logical and physical modelling. This process is a collaborative and open one, and several third party software vendors have decided to use the CDD/Repository information model with their products; others are participating in the top-down review. These include:

- Andersen Consulting
- Information Builders Inc
- Smartstar
- Cognos
- Hypersoft.

5 The COHESION partnership program

COHESION allows third party products to be integrated into the solution and, for many years, Digital has operated a partnership programme with third party software vendors. These have been known as complementary solutions organisations (CSOs). The products are categorised using Digital's own picture of the system development life cycle, as shown in figure 7.6. The table shown in figure 7.7 shows the CSO products at the time of writing.

6 COHESION – conclusion

Digital's CASE strategy is, in many ways, similar to AD/Cycle but has a significant advantage: Digital's repository – CDD/Repository – actually exists, its environment is already in place, and it works.

From a European perspective, however, the choice of third party vendors is very American-biased. Of the CSO products, only Excelerator has a large penetration in Europe. DesignAid has a small

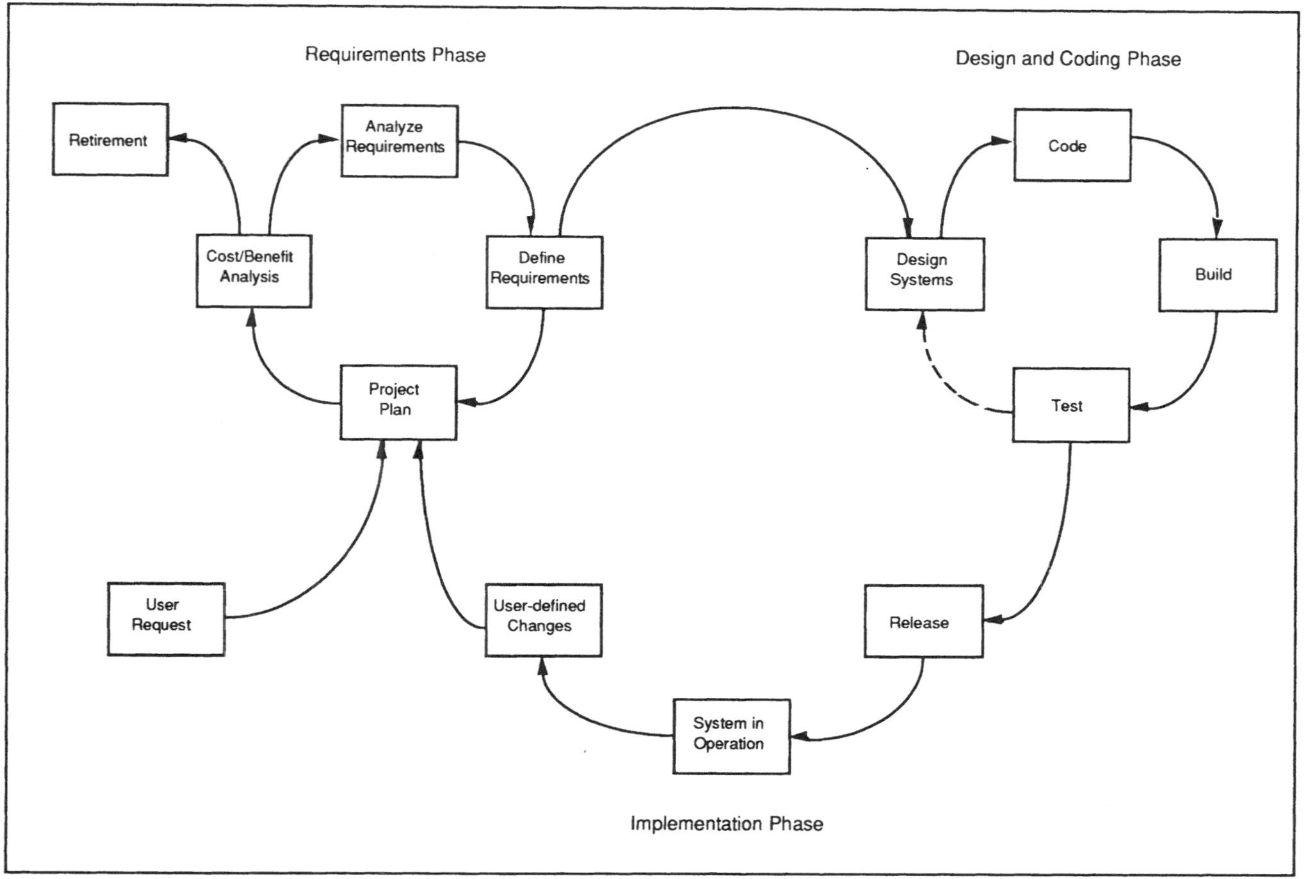

Figure 7.6 Digital's System Development Life Cycle

LIFE CYCLE STAGE	PRODUCT	VENDOR
REQUIREMENTS CAPTURE & ANALYSIS/DESIGN	TEAMWORK DESIGNAID EXCELERATOR SOFTWARE THRU' PICTURES IDEF/LEVERAGE PROMOD STATEMATE TAGS MENTOR CASE VSF ASA & GEODE SPECIF MULTI PRO/SADL SOFTPEN/MIS & RT IDA CONCEPTOR	CADRE TECHNOLOGIES HOSKYNS/CGI INTERSOLV IDE DACOM PROMOD INC I-LOGIC INC TELEDYNE BROWN MENTOR GRAPHICS SYSTEMATICA VERILOG X IGL TECHNOLOGY CAP/SESA SOFTPEN CORP. METSI IBSI/DELTA
IMPLEMENTATION (CODE, 4GL & EXPERT SYSTEMS)	SMARTSTAR POWERHOUSE FOCUS NETRON/CAP CORVISION NATURAL DUNGA NEXPERT OBJECT FOUNDATION ENTERPRISE EXPERT, BUILDER & GENERATOR	SMARTSTAR/SIGNAL COGNOS INFORMATION BUILDERS NETRON INC CORTEX SOFTWARE AG ORDEN NEUTRON DATA SYSTEMS ANDERSEN SOFTWARE CA
IMPLEMENTATION (IPSE)	SABER C SMARTSYSTEM GIE-EMERAUDE	SABER SOFTWARE PROCASE CORP. SYSECA/BULL/EUROSOFT
IMPLEMENTATION (CONFIGURATION MANAGEMENT)	RCS CCC LIFESPAN	ROCHESTER INSTITUTE SOFTOOL YARD SOFTWARE
PROJECT MANAGEMENT	PROJECT 2 QUICKNET ARTEMIS PLANNER METRATECH ACUITY APECS/8000 VVE PROJECT MANAGER	PSDI INC PSDI INC METIER INC. PRODUCTIVITY SOLUTIONS YARD SOFTWARE COMPUTER COGNITION ADP NETWORK SERVICES NATIONAL INFO. SYSTEMS
DOCUMENTATION	INTERLEAF TPS UNIPLEX	INTERLEAF UNIPLEX LTD.

Figure 7.7 Digital CSO Products

number of European users, having been a late entrant into the European CASE market, and both Teamwork and Software thru' Pictures are perceived, in the European market, as real time products: this has limited their general acceptance. Finally, Digital has demonstrated no real ability to market its products successfully against IBM.

So despite its many plus points, the authors feel that NAS and COHESION may not succeed as well in the world marketplace as, technically, they deserve.

7 References

1 Digital (1990), *Digital Vision of Solutions for Software Developers,* June 1990

2 Digital (undated), *Digital CASE Environment: Comprehensive Productivity Solutions,* ECE0089-48/89 09 23 90 0 CTS

3 Digital (undated), *VAXset: VAX Software Engineering Tools,* EC F0340-48/89 09 61 40 0 MKO

4 Digital (undated), *Digital Technical Journal,* ISSN 0898-901X

5 Digital (undated), *VAX/VMS Software: Information Management Handbook,* EB 30904-48/88 1108/36 750 BUO

6 Digital (undated), *Digital's CASE Environment: Comprehensive Productivity Solution Video*

7 Digital (undated), *The Digital COHESION Environment for CASE,* EC-HO465-48/90 06 43 100.0 BUO

8 Digital (1991), *CSO Product Update,* January, March and September 1991

Chapter 8

ICL: CASE Partnerships and OPEN*framework*

1 Introduction

ICL has received considerable attention from the CASE pundits. In September 1988 Infomatics stated that 'ICL led the world by being the first hardware vendor to realise the importance of automating the complete systems development lifecycle'. The Yankee Group reported in June 1988 that 'ICL's work in the CASE/4GL area is highly influential, forming a model that other major 4GL vendors will be forced to follow over the next year'.

ICL's CASE strategy is far from new, having evolved from many years experience with its own tools. Five basic principles provide the cornerstones of this strategy:

- application life cycle – the total portfolio of tools should cover the complete development life cycle
- open systems standards – all elements should conform to internationally agreed or developing standards, such as IRDS, UNIX and X-Windows
- user choice – the framework must allow the user to choose the most appropriate tool, whilst ensuring that information can be exchanged with the dictionary
- collaboration – ICL should collaborate with leading companies whose products complement ICL's

- repository (data dictionary) – this must lie at the heart of all applications developed, formally documenting both the business process and development stages.

2 Information Processing Architecture (IPA)

IPA is ICL's own information processing architecture, on which the whole of the company's range of products – known as the Networked Product Line – are based. IPA incorporates OSI standards, and defines a complete range of standard facilities for interconnection, terminal attachments, application inter-working and system management. IPA currently provides OSI facilities up to the transport (layer 4) level on:

- Series 39 and 2900 Series mainframes
- System 25
- DRS 20/200, DRS 300, DRS 400, DRS 500, DRS 6000, and DRS Intelligent Workstations
- Sun Graphics Workstations
- PC Quattro.

ICL's strategy is to build on IPA's existing OSI level 4 compatibility and incorporate further open standards products and services as they evolve at the higher level. IPA is also being used to ensure compatibility and inter-working with ICL's own products, thus forming a foundation for the CASE strategy.

ICL's product strategy is directed at delivering business solutions to customers within an Open Systems framework. IPA provides conformance across the product range to include the following additional areas:

- application services, such as message handling and file transfer
- application portability
- systems management (for instance, network management)
- information management
- security management.

IPA defines the complete range of open systems standards that all ICL products are required to meet, so as to provide an integrated environment for customers. IPA has to work across the three primary ICL product platforms, supported by MS-DOS, UNIX and VME.

These three platforms, partly overlapping as shown in figure 8.1, are seen by ICL as providing support for corporate (VME), departmental (UNIX) and personal (MS-DOS) computing needs.

ICL's objective is to provide secure and consistent access to applications and services on a network that could involve all three types of platform. This user interface must be built upon industry standards, and has to be transparent to the combinations of technologies engaged, while applying equally to all types of users. Figure 8.2 illustrates this approach.

The current phase of ICL's programme includes tools to build personal and shared integrated environments for end users. Major developments in access standards have also seen ICL introducing the following graphical user interface (GUI) options:

- human computer interface (HCI) standards for UNIX follow those defined by X/Open and UNIX International – on the DRS 6000 ICL has provided support for X.11, allowing the adoption of both Open Look and OSF/Motif
- DRS models that run MS-DOS can support Windows 3, and ICL has provided a product called PC-Connect that allows users to access both MS-DOS and UNIX applications in a consistent way
- Presentation Manager support is provided for DRS models running OS/2.

3 OPEN*framework*

ICL reaffirmed its commitment to open systems in June 1991 with the release of OPEN*framework* – a blueprint for distributed computing operations. This architecture is the company's vision for fully integrated information systems, based on open systems standards and distributed, object-oriented computing. It addresses the needs of an organisation from four different perspectives:

- enterprise management – those responsible for ensuring that IT supports business goals, and that systems respond to business dynamics
- users – those using information systems in support of the organisation's goals

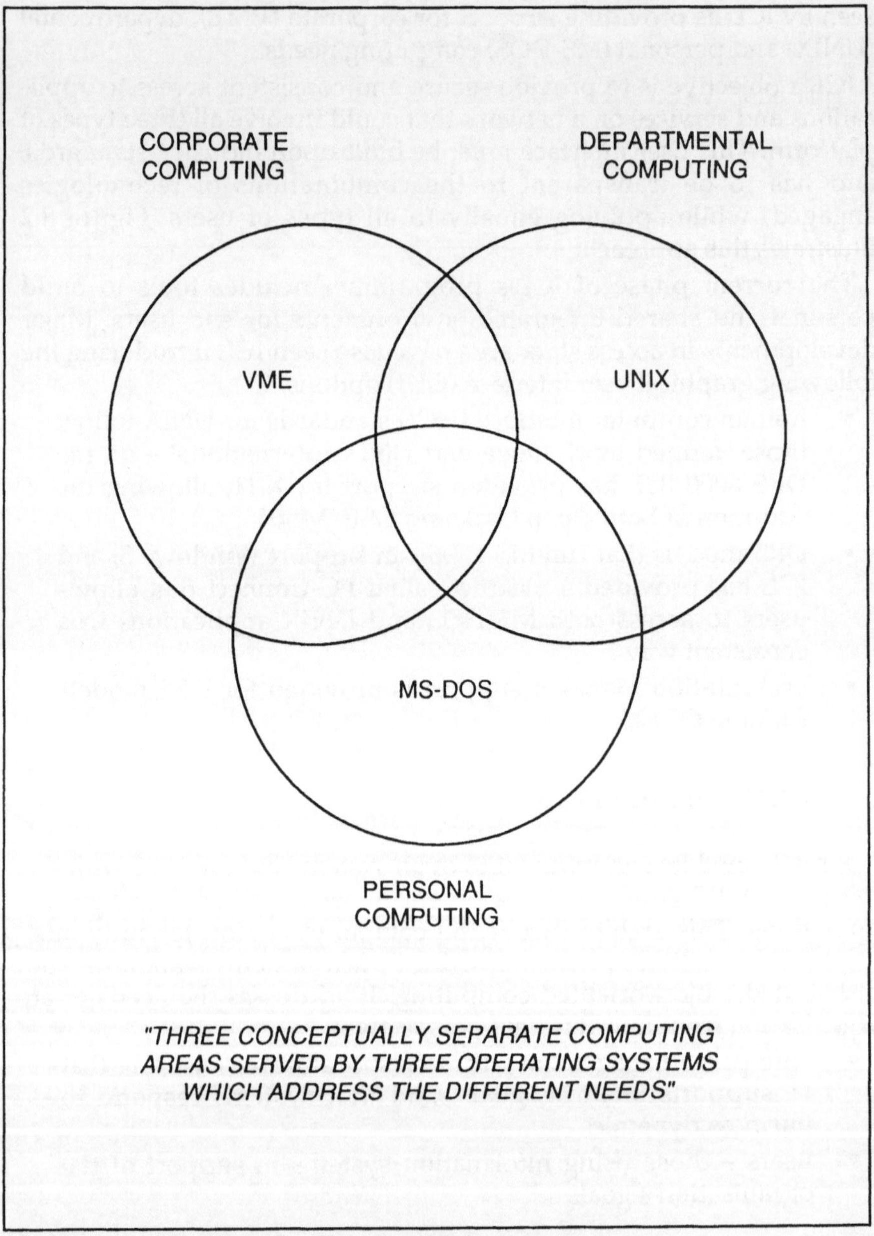

Figure 8.1 The Three Overlapping Environments

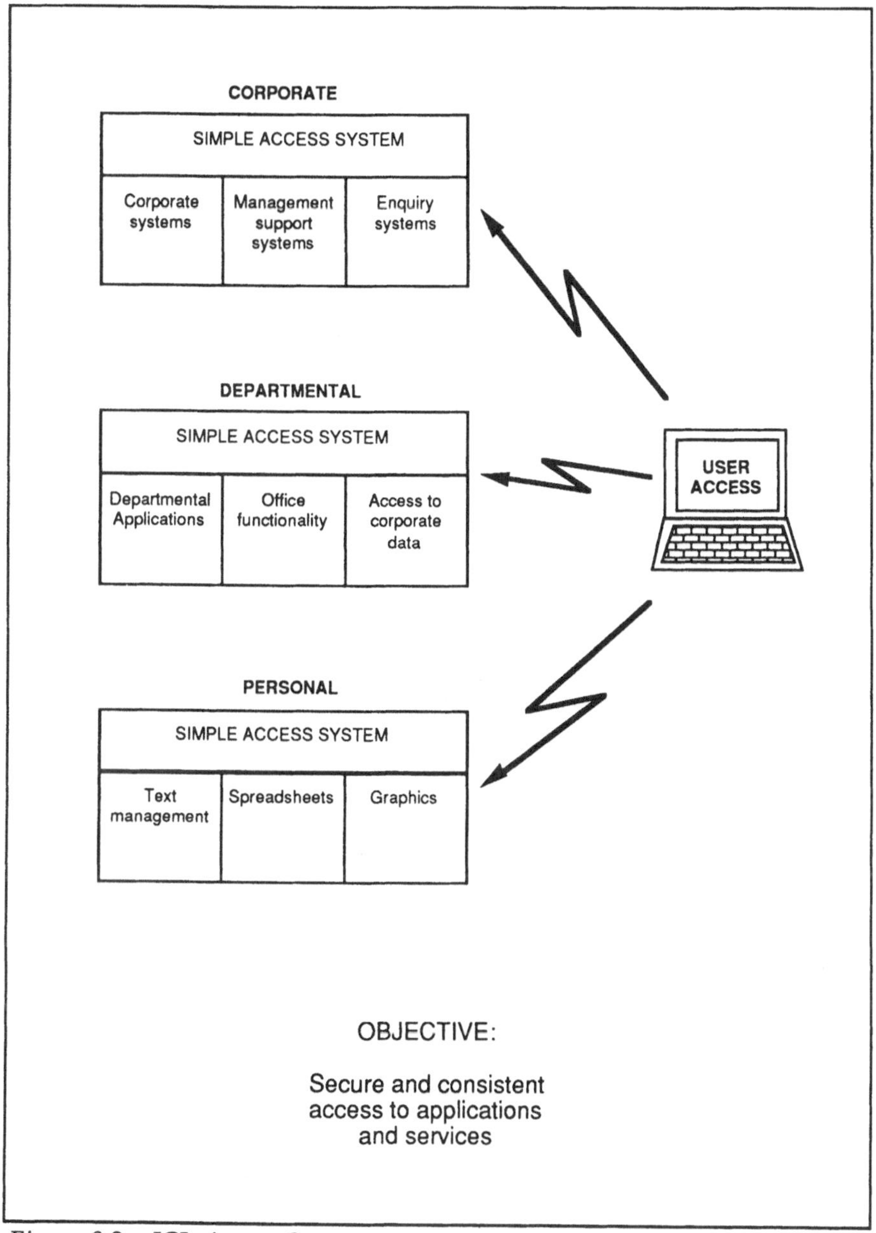

Figure 8.2 ICL Access Systems

- application developers – those who select, develop or procure, test and integrate applications for an organisation
- service providers – those who administer and control the information systems infrastructure, and optimise the value of the organisation's investment in IT.

OPEN*framework* includes rules for assuring the quality of conforming products. Service levels are set for:

- availability – a measure of whether the system is there when needed, whether the required service is provided, and whether it meets agreed service levels
- usability – a measure of how effectively specified users are able to carry out specified tasks in a particular environment,
- performance – a measure of system throughput and response
- security – a measure of the ability of a system to withstand unauthorised access and to identify users with their actions
- potential for change – a measure of the ability of an information system to evolve and change in response to new requirements, and of its ability to adapt and re-use components.

Individual products within OPEN*framework* are tested for conformance to open system standards at ICL's System Integration and Validation Centre in Bracknell (UK). As a result of this validation activity, ICL offers a guaranteed inter-working set of open systems products. An integration guide was made available in autumn of 1991.

OPEN*framework* has a simple layered structure comprising eight interlinked elements, as shown in figure 8.3. These are:

- user interface
- application architecture
- distributed application services
- information management
- application development
- systems management
- networking services
- platforms.

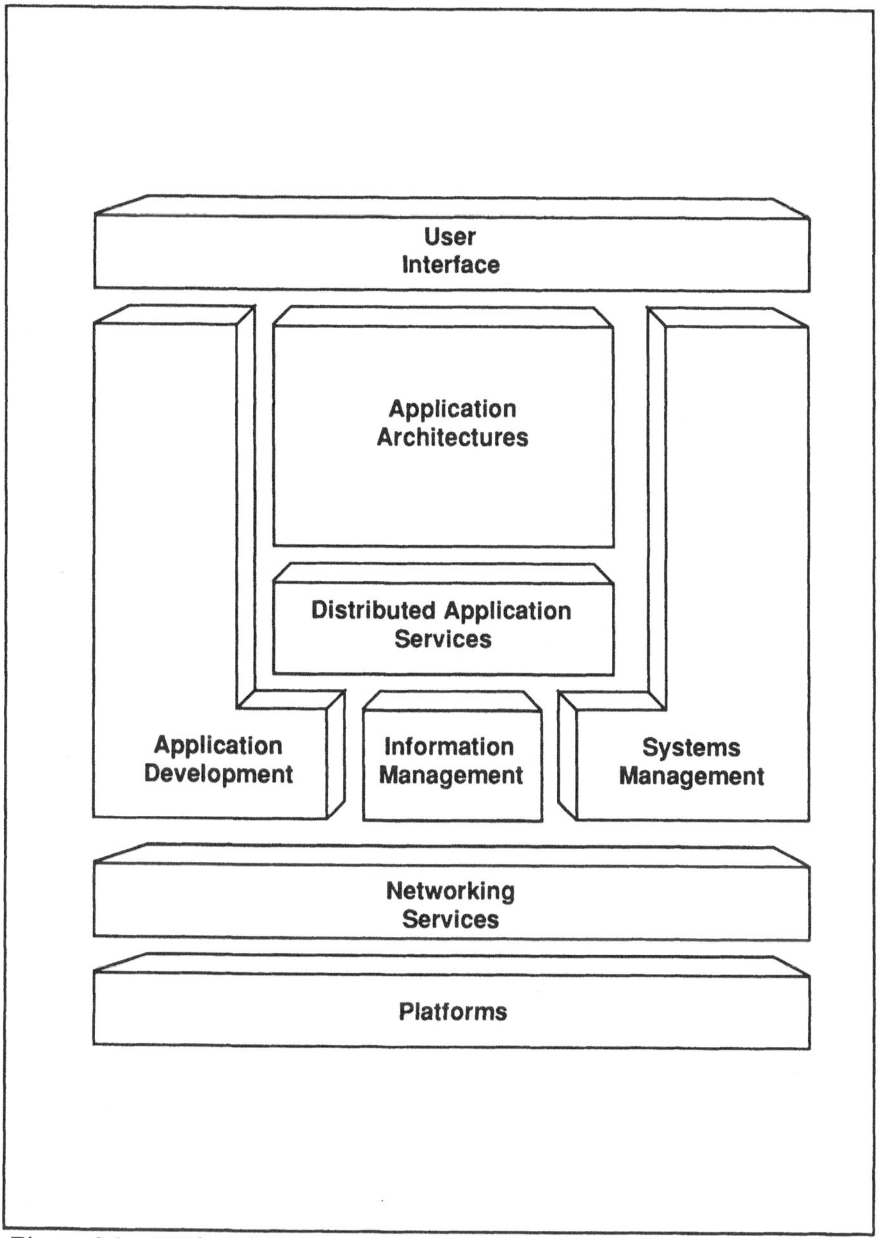

Figure 8.3 ICL's OPENframework

The user interface makes the information available to a wide spectrum of users. Each will be presented with a consistent windows-style interface, irrespective of what application is running, or whether it is running on a PC, a UNIX server or a VME mainframe. As well as the X-Windows standard, popular graphical user interface products such as Windows 3, OPEN LOOK and OSF/MOTIF are included. Character interfaces are catered for by the de facto VT 220 and 3270 standards.

The application architecture will initially be supported by ICL's OFFICEPOWER office automation system. This will be followed by line-of-business applications in retailing and other specific markets targeted by ICL. Application architectures can be built within the overall framework to meet specific needs, and ICL is expecting organisations and software vendors to develop their own application architectures to complement those provided by ICL.

Distributed application services allow transaction processing applications to be distributed across different platforms. OSI/TP is one of the key standards in this area. For UNIX systems, TUXEDO will be ICL's strategic product. Additionally, OPEN*framework* will provide inter-operability with OSF's distributed computing environment, and will support the OSI/RPC (remote procedure call) standard now being defined.

The information management element is where information is modelled, stored, retrieved and interchanged. CODASYL (IDMS/X), relational (INGRES, Oracle, Informix) and object-oriented (ODAC) database management systems are currently supported.

The application development element contains CASE tools, provided both by ICL and by independent software vendors, to support the systems development lifecycle. An open dictionary will be provided as the central repository for the intellectual property of the users, acting as the focus for the integration of tools and data. Sample tools include: DDS, QUICKBUILD, ADD, Westmount ISEE and COBOL.

Systems management facilities are provided by the open systems management centre (OSMC). This comprises hardware, software, user interfaces and network infrastructure, and has been designed to handle, from central points, the operational control of multi-vendor distributed systems. By using OSMC, it is possible to carry out fault diagnostics, reload software on remote servers and allocate additional file store to any area of the network. OSMC has the ability to log and progress all planned enhancements or repairs to the network, and all service requests from users.

Networking services are provided to support information interchange between computer platforms and workstations. Inter-working services include virtual terminal protocol, message handling (X.400), directory services (X.500), and File Transfer Services such as FTAM. These are supported by wide and local area network platform interconnection services such as X.25, SNA, SDLC, OSLAN and TCP/IP.

Platforms include workstations and PCs supported by departmental and corporate servers, controlled by operating systems such as MS-DOS, UNIX and VME, as well as others that conform to accepted standards.

4 ICL CASE components

There are a number of aspects to developing and maintaining software under the ICL CASE umbrella. Figure 8.4 illustrates these components.

Methods provide the structured and disciplined approach to developing applications. ICL CASE does not stipulate that any particular method must be used, but methods supported include :

- SSADM
- Information Engineering
- Merise
- strategic information systems planning
- project management.

ICL has been at the forefront of dictionary technology since the late 1970s. They have had a close involvement with the BCS Data Dictionary Working Party, and now with the BSI Information Resource Dictionary Systems (IRDS) Rapporteur Group. Their CASE strategy is based around their dictionary – DDS. However DDS suffers from three major drawbacks:

- the reliance on IDMS/X
- poor reporting facilities
- interfacing difficulties.

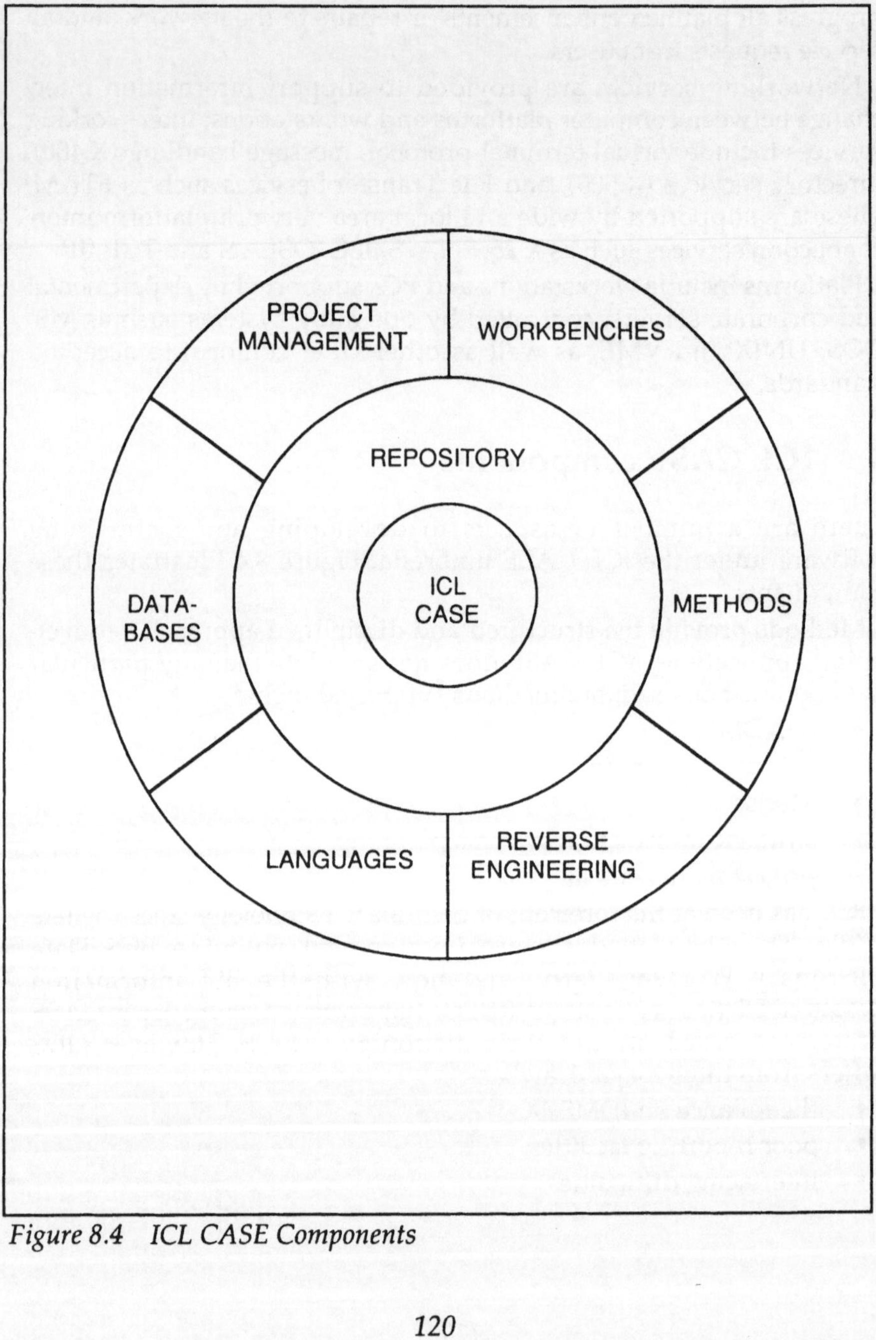

Figure 8.4 ICL CASE Components

The database engine is IDMS/X, a CODASYL DBMS which, when DDS was designed, required significant compromises in the meta-model in order that it could fit within the performance limits of the database engine. Despite subsequent improvements in the database engine's performance, no alterations to the physical database structure of DDS were made. This has made the dictionary difficult to extend. In addition the commonly-accepted. requirement for the 1990s is the use of SQL as the user manipulation language.

DDS reporting is complex and less than user friendly. No efforts have been made by ICL to improve the output facilities. Since 1987, various third party vendors have been producing products to fill this gap. Examples are DDS/Report from Gresham, REPMAN from COBRA Systems and LADD from Loud and Bows.

Because of the complex nature of the DDS structure, interfacing to other products has not been an easy task. ICL has not, in the past, encouraged third party vendors to interface to DDS, which can therefore be viewed as a closed product.

The 1989 strategy saw interfacing problem tackled, but the DBMS-dependency made more complex for ICL users through the addition of an alternative dictionary product. ICL's database strategy for the 90s is linked to INGRES, and the company has entered into an agreement with Sema to use ADD, an INGRES-based dictionary, on ICL's UNIX machines. This product has links already to LBMS' Automate Plus and System Engineer products, and the intention is to produce a link between ADD and DDS to allow information to be exchanged, as shown in Figure 8.5.

ICL's CASE tools can operate with both IDMS/X and INGRES dbms products. Known as workbenches, they provide the developer with a working environment, and are used for a variety of specific tasks supporting one or more stages of the life cycle. ICL offers two products:

- Quickbuild WorkBench – providing facilities for using Applications Master (ICL's 4GL product)
- Programmer's WorkBench – allowing users to create and modify COBOL programmes.

A number of programming languages can be used including COBOL, C and SQL as well as 4GLs. ICL's offerings for 4GLs are Application Master for VME and INGRES/4GL for INGRES applications.

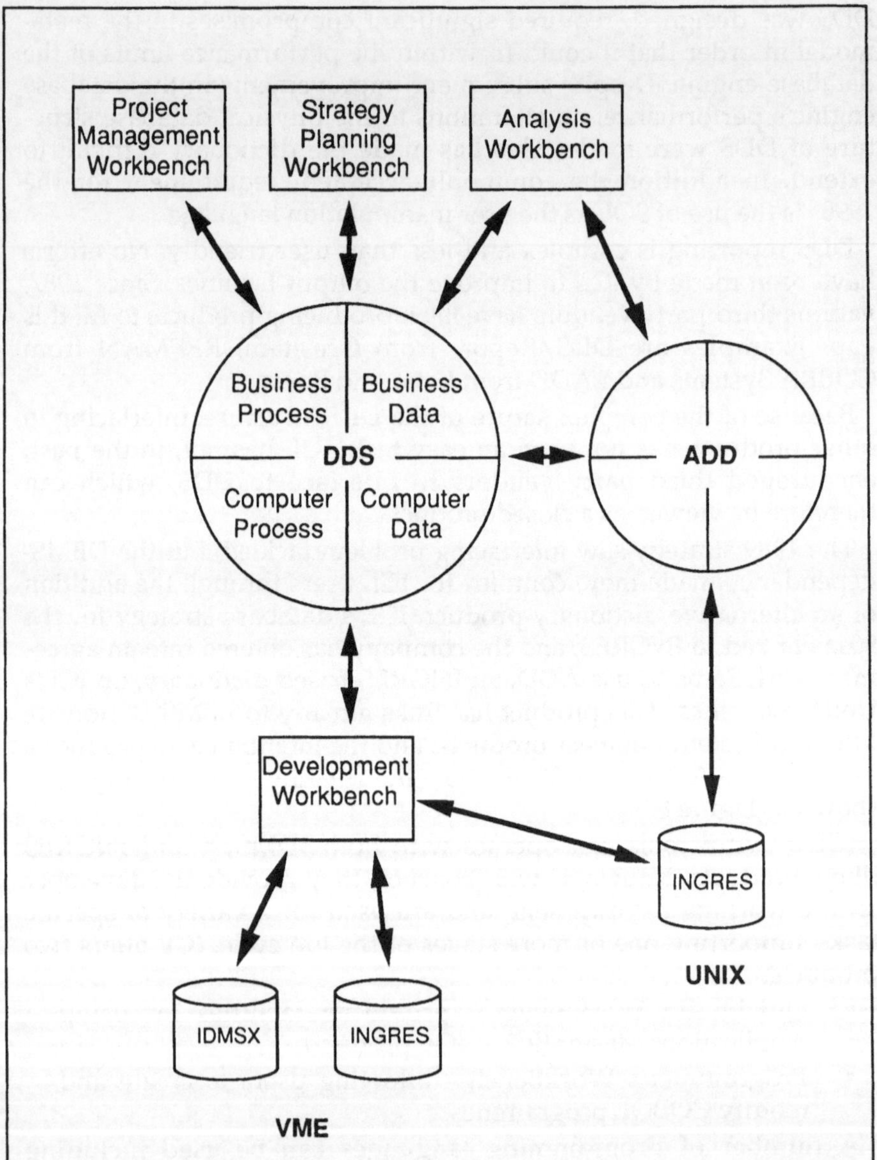

Figure 8.5 Relationship Between ICL's Dictionaries and Workbench Tools

Project management tools provide for better management of a complete development project. Hoskyns' PMW is the CASE partner product selected for linking into the ICL CASE strategy.

Reverse engineering and re-engineering are included to allow existing systems to be backtracked and documented before a rewrite. ICL originally chose PATHVU and RETROFIT as its CASE partner products. Unfortunately, the choice was made just as KPMG were winding up their Catalyst division, selling the products to XA Systems. Design Recovery from Intersolv and REVENG from LBMS are the current chosen tools. In addition, Delta Software are working on integrating their Aemelio product into the ICL CASE Environment.

Figure 8.6 shows the ICL CASE component products against the supported life cycle.

5 ICL's Open Dictionary

ICL has decided to build a new dictionary product for the 1990s, based on the IRDS standards, as a key requirement of the ICL open systems framework for inter-working distributed corporate, departmental and personal services. This is called, at present, the Open Dictionary. The main requirements to be met by this new product include:

- conformance to standards – the product conforms to the IRDS frame work standard (ISO10027), and in addition is flexible enough to conform to interface standards such as PCTE and CDIF, as well as adopting an object-oriented architecture conforming to OMG's object management architecture
- compatibility – the new product must protect the investment made by customers in DDS, both in terms of dictionary data and in software and working practices
- architectural integration: the new product is part of OPEN*framework* as well as of ICL's application development technology
- complex objects – the product must be rich in functionality, especially in the areas of object naming, version control and query support, with support for complex information including graphics and word processed text

		BUSINESS PLANNING	SYSTEMS ANALYSIS	SYSTEMS DESIGN	DEVELOPMENT	IMPLEMENTATION	MAINTENANCE
METHODOLOGY	SSADM	PA TETRARCH LBMS STRATEGY PLANNER	AUTOMATE PLUS+ EXCELERATOR+	AUTOMATE PLUS+ EXCELERATOR+			
	IE	IEW PLANNING WORKSTATION	IEW ANALYSIS WORKSTATION	IEW DESIGN WORKSTATION			
	OTHER	PA TETRARCH	EXCELERATOR+ AUTOMATE PLUS+	EXCELERATOR+ AUTOMATE PLUS+			
DICTIONARY		ICL DATA DICTIONARY SYSTEM (DDS) ON VME SEMA GROUP ADVANCED DATA DICTONARY (ADD) ON UNIX					
PRODUCTIVITY	DATABASE			IDMSX-VME+ INGRES-VME+ INGRES-VMX+	IDMSX+ INGRES+	IDMSX+ INGRES+	IDMSX+ INGRES+
	DEVELOPMENT			QUICKBUILD WB+	QUICKBUILD WB+ INGRES FORMS+ PROGRAMMERS WB+	QUICKBUILD WB+ INGRES FORMS+ PROGRAMMERS WB+	QUICKBUILD WB+ INGRES FORMS+ PROGRAMMERS WB+
PROJECT MANAGEMENT		HOSKYNS PROJECT MANAGER WORKBENCH (PMW)					

KEY

Name + - Product has already been linked to DDS

Figure 8.6 ICL CASE Component Products

- active – the dictionary must be able to access external software and/or data so as to effect data transformations with minimal coding effort
- quality attributes – the product must satisfy the quality conditions for OPEN*framework*.

The architecture of the Open Dictionary (COSMOS program) is shown in Figure 8.7.

The principal components are:

- a user interface system, based on standard graphical user interface technology, and conforming to accepted standards for look and feel
- an object-oriented database system (codenamed Raleigh), for managing the storage and retrieval of complex objects
- a software bus, the object request handler, provided as an integral component of Raleigh but which also enables access to external objects via OPEN*framework* distributed applications services
- a set of servers providing additional functionality, implemented as objects within Raleigh – for instance a text analyser, a graphical server and a document server
- interface modules, providing a variety of software interfaces to the dictionary
- the dictionary schema, defining the object types that can be stored in the dictionary
- other tools, which are not part of the dictionary system itself.

6 ICL CASE partners

The CASE partners programme consists of a series of collaborations, in which leading CASE suppliers work in partnership with ICL and ICL's customers. The current CASE partners are :

- PA Consulting Group
- LBMS
- Softlab
- Ernst & Young (IES)
- Intersolv UK
- NCC

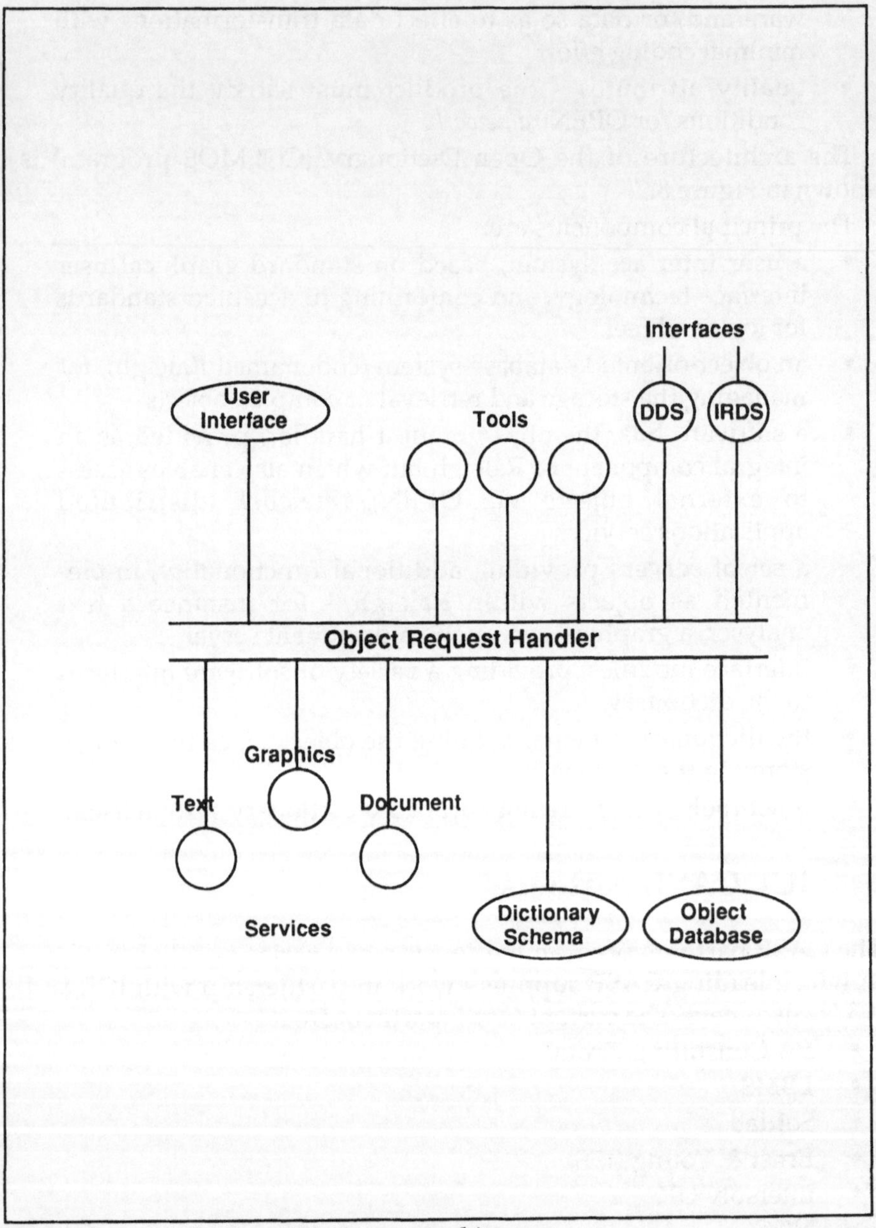

Figure 8.7 ICL's Open Dictionary Architecture

- Hoskyns
- Software One
- SQL Systems International
- Westmount.

PA Consulting Group supplies TETRARCH, which provides support for strategic information systems planning, and is being developed to support an interchange with DDS.

LBMS' Strategic Planner, Automate Plus and Systems Engineer PC products are included. Automate Plus and Systems Engineer already have interfaces to DDS, and Systems Engineer has an interface to ADD. Strategic Planner has an existing interface to Automate Plus. The partnership agreement also includes REVENG, the LBMS toolset and services handling reverse and re-engineering. Currently REVENG can capture information from COBOL and other systems and can exchange information with DDS through an ASCII file.

Softlab's Maestro II (see chapter 17) supports methodologies such as SSADM, LSDM/Systems Engineering and Merise. The original product, Maestro, provided terminal emulation support and a COBOL development environment for ICL VME. Maestro II can be delivered on DRS 6000 boxes and a link to DDS is being developed.

Ernst & Young (IES) supplies the IEW product set. This, and in particular the Analysis Workstation component, form a partnership product to support customers using Information Engineering methodologies. The Analysis Workstation has a two-way interface with DDS.

The agreement with Intersolv UK includes not only Excelerator but also PC PRISM and DESIGN RECOVERY. Excelerator supports a variety of methodologies including SSADM and has full links not only with DDS but also with another CASE partner tool, Hoskyns' PMW. PC PRISM is a strategic planning support tool which interfaces with Excelerator. Design Recovery is a PC-based reverse engineering tool that uses the Excelerator dictionary to store captured information.

The National Computing Centre's ASSET is a multi-user LAN-based front end CASE tool that supports SSADM Version 3, stages 1 to 3 only. An interface to DDS is planned.

Hoskyns' PMW is the selected project management workbench tool for the ICL CASE environment. The company also supplies BRIDGE, a resource estimating tool that works with PMW. PMW interfaces with a number of front end CASE tools, and links with DDS are planned.

Software One's EXCHANGE is a product that provides the capability of linking different CASE tools, allowing a bi-directional exchange of information from one tool to another.

SQL Systems International supplies PCMS, a configuration management system that runs under UNIX. It provides support for change control , release control, build control and version control. There are special networking features to facilitate connections between UNIX and VME environments.

Westmount's ISEE is a front end CASE tool which is integrated with a number of RDBMS products including INGRES. It provides an applications development environment for analysis and design, using Yourdon Structured Method (YSM) techniques, through to the automatic generation of databases and programs. A reverse engineering facility for rdbms tables is included.

7 ICL CASE strategy – conclusions

ICL's stated intention is to continue to expand and enhance the strategy over the next few years. This was announced before the company's takeover in 1990 by Fujitsu, and also predates the takeover of INGRES by the ASK Corporation. In our view, the introduction of ADD weakens ICL's approach, as it introduces a dual repository standard. If ICL does deliver a replacement repository for the 90s, then that could resolve this issue – so long as it provided a means of migrating from ADD and DDS.

ICL has also talked about seeking collaboration with other vendors in the area of quality management. No announcement has been made.

Integration is planned between office support (OFFICEPOWER) and the CASE environment.

ICL has devised a slogan for its CASE Strategy :

C is for Collaboration with the best third party vendors.
A is for Adopting standards such as IRDS, SQL, C, SSADM, Information Engineering and UNIX.
S is for Strengthening their current products.
E is for Enhancing products through integration.

8 References

1 ICL (1989), *ICL Inform*, Issue No. 31, November 1989

2 ICL (date unknown), *CASE brochure*, CS 3201 (36783)

3 ICL (date unknown), *Partners in Progress*

4 ICL (1991), *ICL Inform*, Issue No.48, Supplement, June 1991

5 ICL (date unknown), *Application Development: The Open Dictionary*, Technical Overview, SE-K-COS-2, Issue 2.0

Chapter 9

UNiSYS: Integrated Information Architecture

1 Introduction

During 1989 and 1990, UNiSYS conducted a strategic exercise, intended to define a technical architecture which could incorporate existing and future hardware and software products. This architecture is termed the Integrated Information Environment (IEE). It has been designed to cater for the multi-vendor world, and its approach to inter-operability is based on open system standards and extensions to these standards where they provide value-added functionality.

This chapter looks, first of all, at IEE as a whole, before examining the CASE related services inside it.

2 IEE architectural overview

From a functional standpoint IEE consists of five sets of services, as shown in Figure 9.1:

- systems connectivity services (SCS) provide for the transmission of data, text, graphics, image and voice information, and satisfy the communications cost/performance requirements of the business
- distributed systems services (DSS) support the information distribution and resource planning requirements of the company, both internally and externally

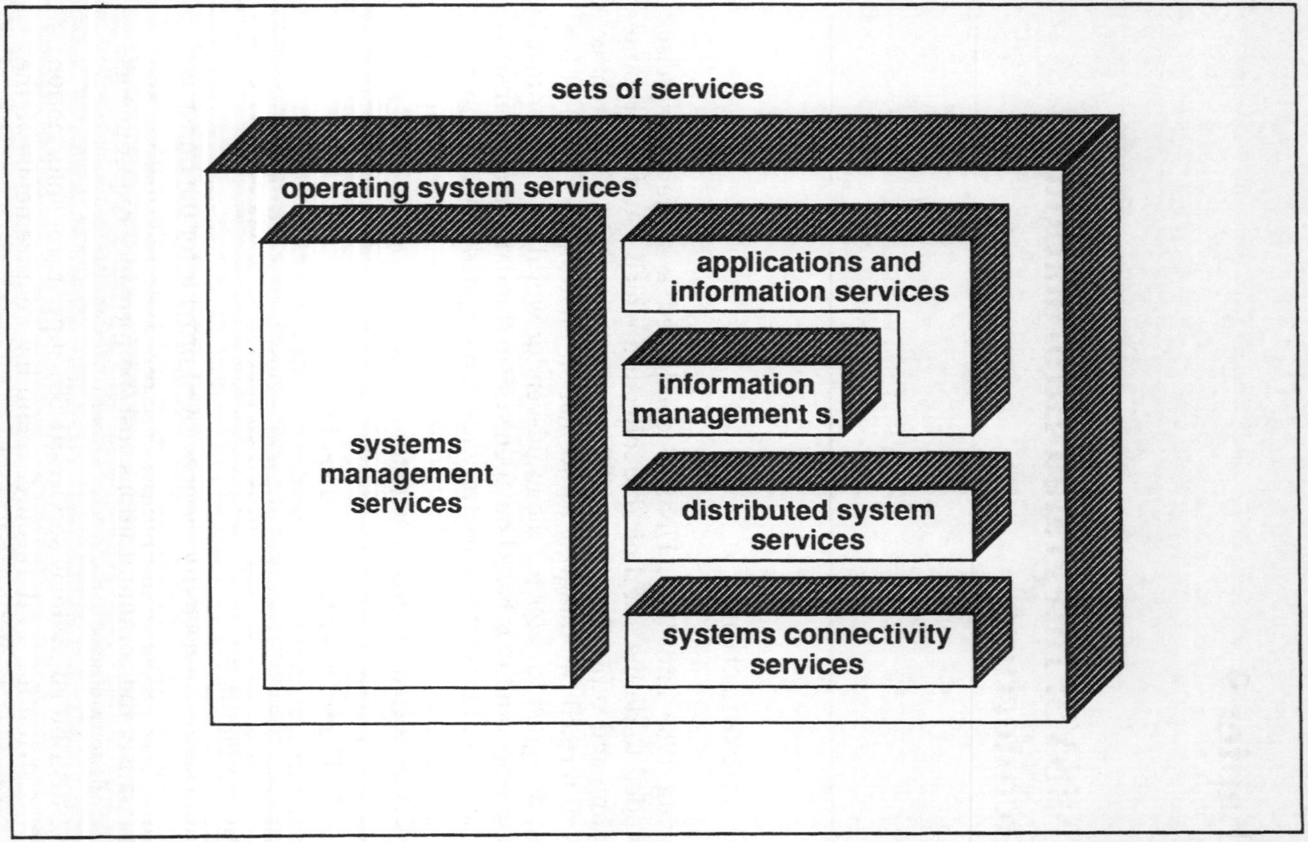

Figure 9.1 IEE Services

- information management services (IMS) provide for the secure storage, retrieval and presentation of information throughout a company
- application and information services (A&IS) allow the design, integration, implementation and maintenance of applications that respond to business needs, whilst improving productivity and effectiveness of individuals
- systems management services (SMS) enable a company to define, implement, and enforce operational policies and procedures that control, monitor and safeguard its information system assets.

The service sets are grouped by UNiSYS into three subsystems:

- network, consisting of SCS and DSS
- application, consisting of IMS and A&IS
- management, consisting of just SMS.

Each service is categorised into one of the following classes of capability:

- open services, defining the capability, interfaces and products required by industry-defined standards identified in the architecture
- premium services, providing either added value extensions beyond the open class of capability, or unique capabilities not addressed by standards
- complementary services, providing the capability to permit coexistence with targets including products adhering to de facto standards, or to other vendor architectures that emphasise support of proprietary standards.

These types of platform are integrated into an information network, on which are based:

- information hubs, nodes that act as servers to the rest of the information network
- servers, network and regional nodes that compose a unified operation environment for the information hubs
- workstations/workgroups, nodes that provide individual end users with gateways to the information network.

3 The application subsystem

The application subsystem is the most visible part of any system in an IEE-based company. It controls the storage of information, the applications that use these databases, and the common end user interface (EUI). The EUI provides access to applications and information in a consistent 'look and feel' way.

The information management services must provide for the secure storage, retrieval and sharing of business information throughout a company. To meet this purpose, the IMS provides:

- tools for modelling both entities, about which the company maintains information, and processes, by which the company acts on the entities
- support for grouping these entities logically into databases of related information
- the functions necessary to ensure that databases are readily available and able to perform at required levels
- the services used to link mission-critical applications and end users with the information they require
- a common store for the deposit and retrieval of development information by tools and procedures
- a common store for information about all aspects of the production environment.

Figure 9.2 shows the functional interface that the IMS will support.

The two major components of IMS which respond to these requirements are:

- database management systems (DMS)
- UNiSYS repository.

The database management systems (DMS) provide for the management of data in both logical and physical form, and supply the means for database definition, administration and access. The UNiSYS architecture supports relational database management systems now, and targets object oriented databases for the future. The relational model offers flexibility, data independence and a standard interface – SQL.

The semantic/object oriented model will offer the performance of the hierarchical model, the flexibility and data independence of the

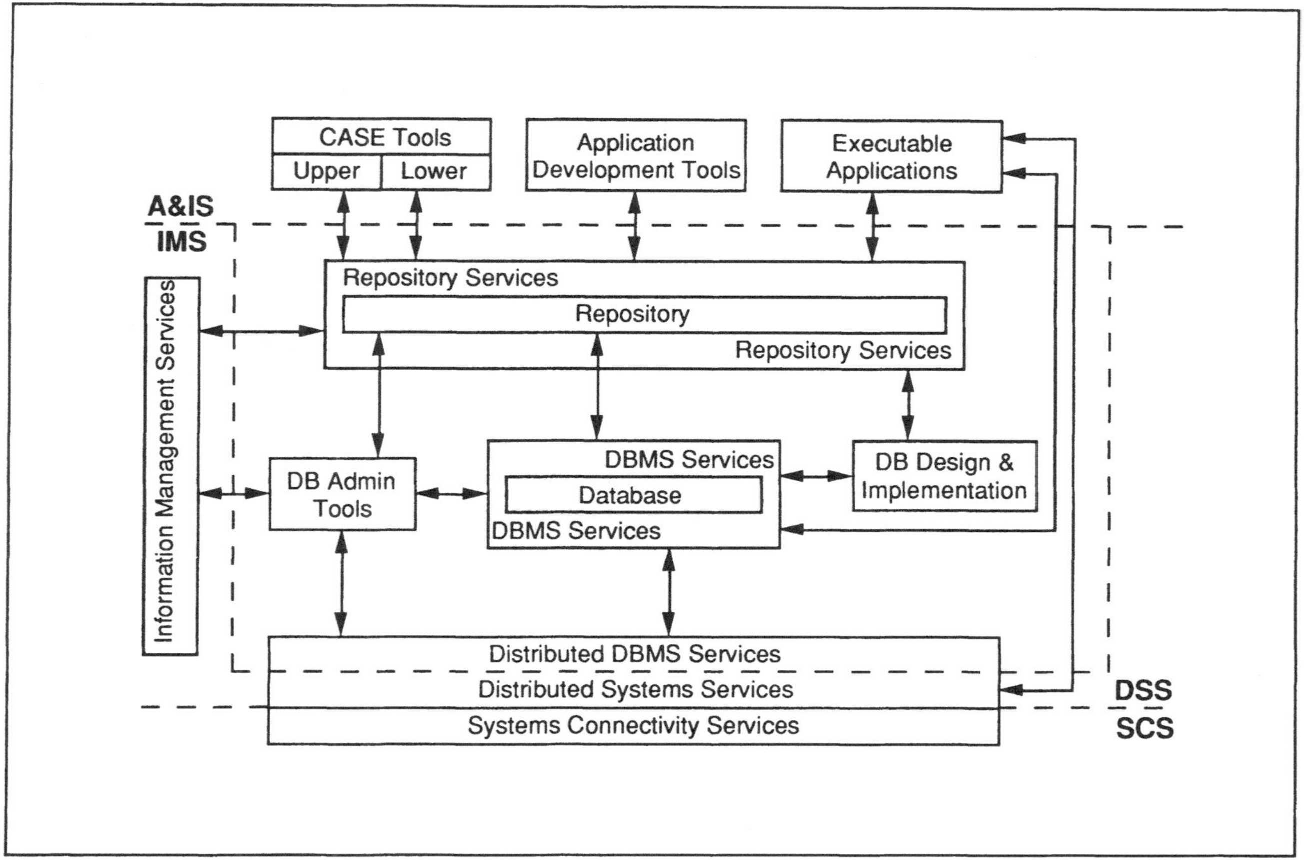

Figure 9.2 IIE IMS Functional Interfaces

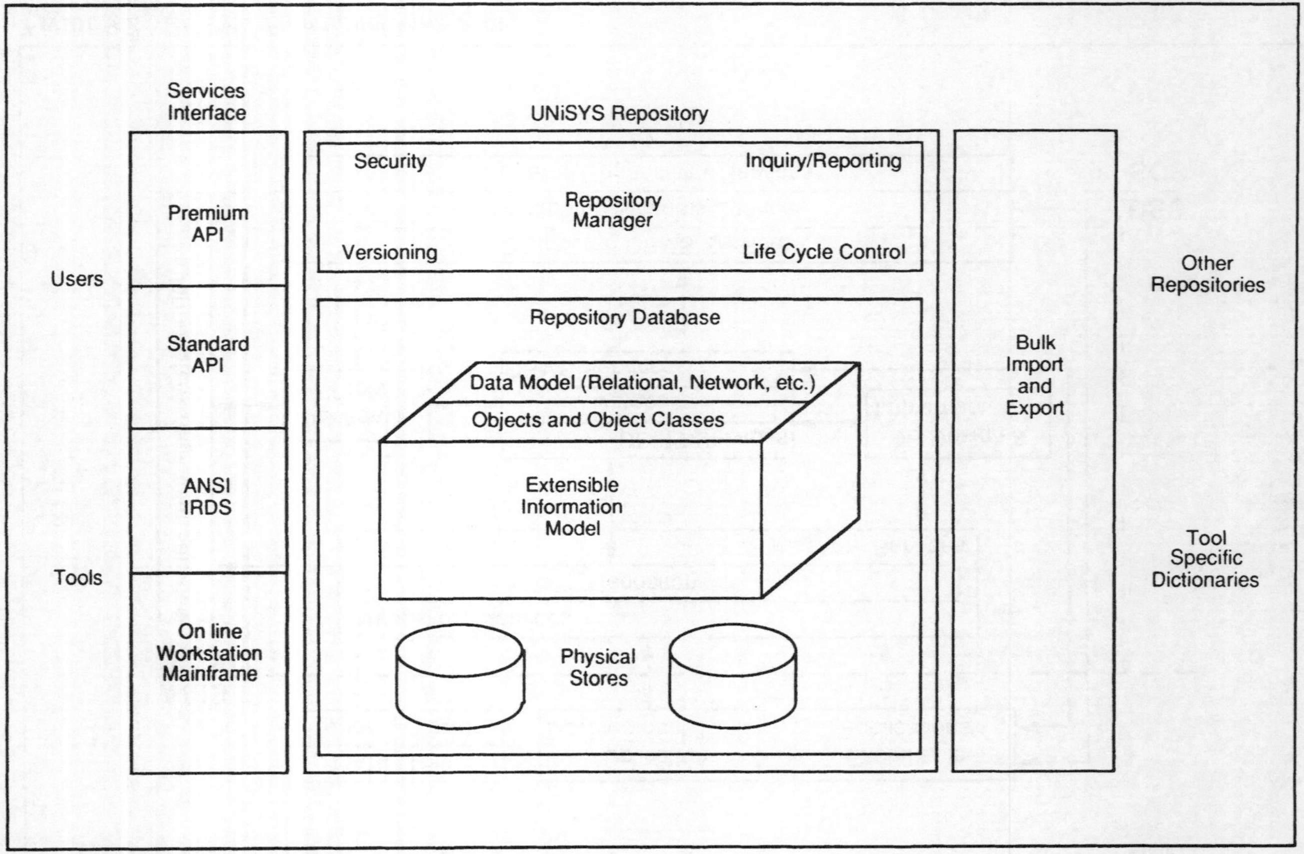

Figure 9.3 The UNiSYS Repository Structure

relational model, and extensive support for maintaining integrity and boosting the productivity of both application developers and users.

The UNiSYS repository will form the backbone for the operation and maintenance of applications using both 3GL and CASE/4GL products. It consists of:

- the repository database, the actual store for the information controlled by the repository
- the repository manager, providing features for controlling and utilising the repository database, including:
  - versioning and life cycle management
  - standard reporting
  - security
  - other administrative functions
- the repository services interface, enabling the repository to interact with the outside world through application programming interfaces (APIs), providing an interface for online interactive access, and supporting ANSI standard IRDS calls.

The repository will facilitate the transmission of information across the application life cycle. Systems management services (SMS) will use the repository and the database administration functions in accomplishing its assigned rôle (see below). LMS uses the distributed systems services (DSS) and systems connectivity services (SCS) to link it with the rest of the enterprise's information resources. Figure 9.3 shows the UNiSYS repository structure.

The application and information services are designed to enable the design, integration, implementation and maintenance of flexible applications that respond to business needs and improve productivity and effectiveness. It consists of the following:

- user interface service, providing access to both IS applications and end-user information services
- development services for both IS and end-user applications
- production services for executing IS applications
- information services for meeting end-user requirements.

The development services component of A&IS supports the application development process, mapping it onto the following system development life cycle:

- enterprise analysis, which supports specification and modelling of an enterprise business infrastructure to construct the enterprise information model
- systems analysis, which addresses the specification and modelling of application systems that support, or are needed to support, the enterprise information model
- systems design, which uses specifications and models generated by systems analysis to enable system designers to determine how an application must be structured, at the system level, to satisfy identified requirements
- detailed design, which uses the specifications generated during the system design phase to enable application designers and programmers to develop detailed specifications for each of the application's modules, databases, files and reports
- coding and generation, which permits application programmers either to code or automatically generate the application
- integration and testing, which assists in integrating the proper application modules, and in testing the application system
- maintenance, which supports post-production reporting and debugging of problems, as well as updating of the application.

Figure 9.4 portrays the A&IS development services model, and figure 9.5 shows its components.

The first three phases of the UNiSYS systems development life cycle – enterprise analysis, systems analysis and systems design – are addressed by upper CASE services. They are workstation-based, graphics oriented tools that improve productivity and overall application quality. The last four phase – detailed design, coding and generation, integration and testing, and maintenance – are addressed by lower CASE services. These consist of detailed design tools, editors, languages and compilers, development tool kits, testing and debugging aids, and generators. The models produced by upper CASE tools are stored in the UNiSYS repository for use by the lower CASE tools when generating complete applications. Extending across all phases are the cross-life cycle services, which include:

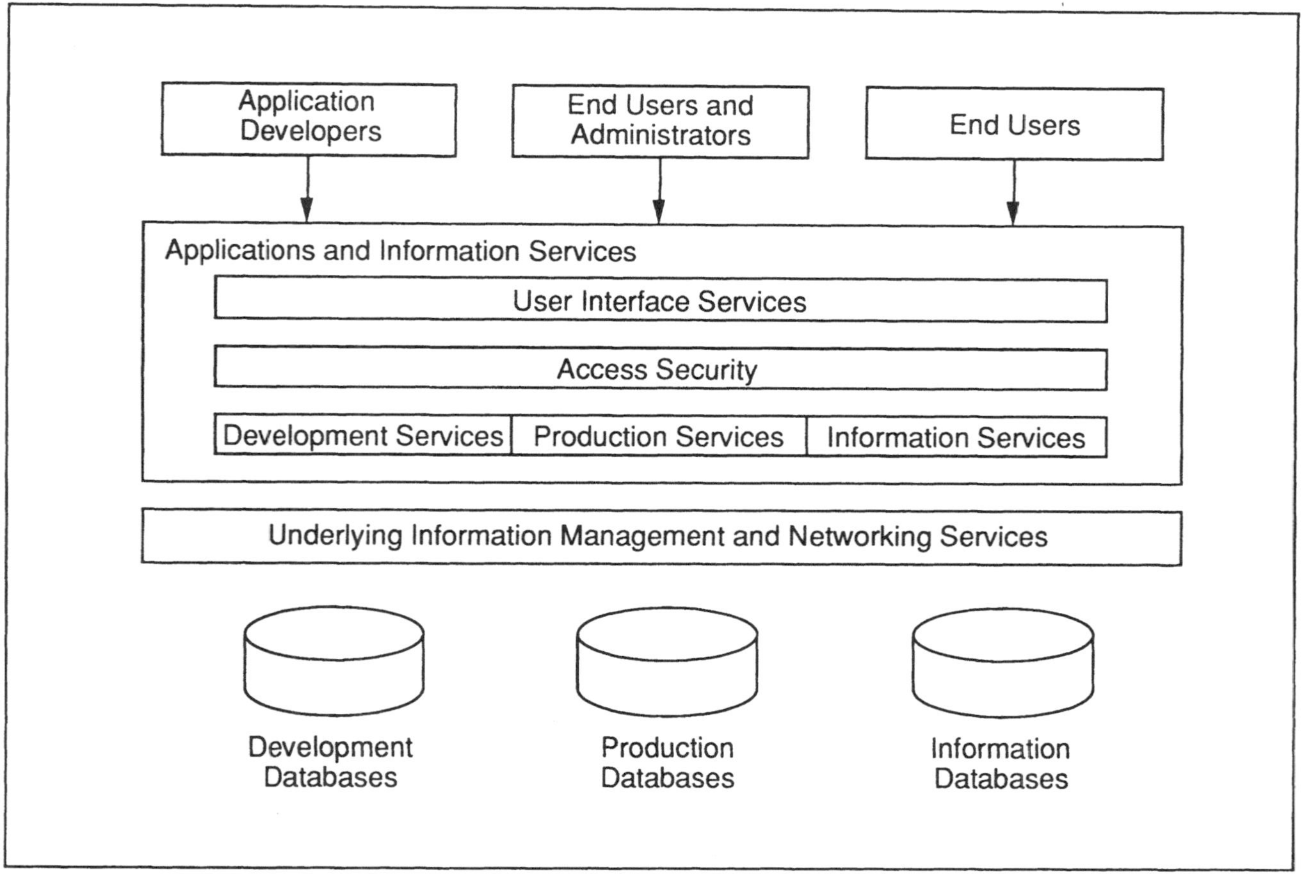

Figure 9.4 The UNiSYS Applications & Information Services Model

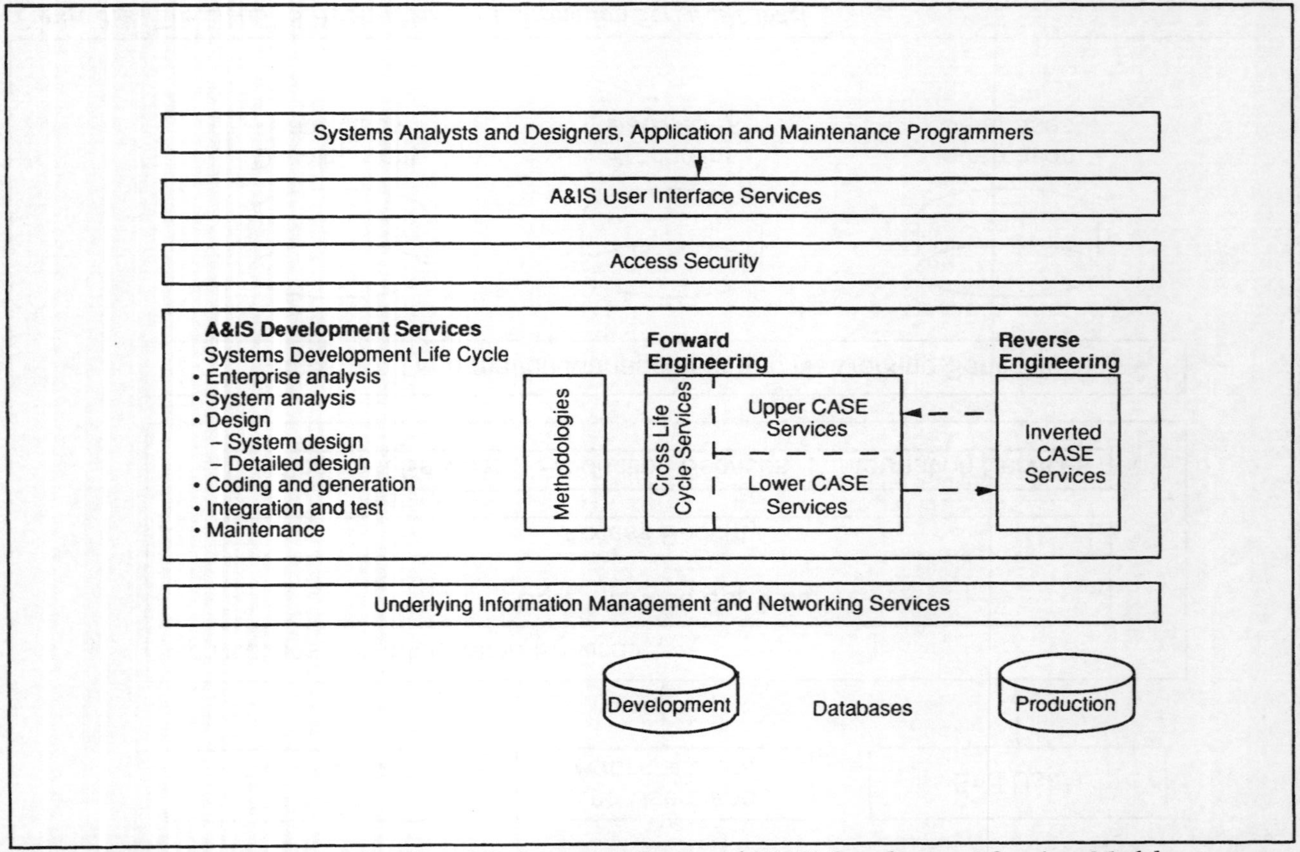

Figure 9.5 The UNiSYS Applications & Information Services (A&IS) Development Services Model

- project planning, to facilitate the definition of deliverables, activities, detailed work breakdowns and work assignments associated with a project
- resource estimating, to aid in determining manpower requirements in each project phase
- activity scheduling, to assist in defining the size of each activity, as well as inter-dependencies between activities
- project reporting, which maps the current status of project deliverables against the project plan
- impact analysis, which assists in replanning necessitated by unforeseen circumstances.

4 UNiSYS' advanced solutions development (ASD) framework

The ASD framework was announced by UNiSYS during 1991 as a further development of IIE. It lays down the software foundation for creating applications solutions for information networks in an integrated, open systems environment. As such, ASD fulfils two key objectives of IIE:

- it provides for the secure storage, retrieval and sharing of core business and end-user information throughout an organisation
- it enables the design, integration, implementation and maintenance of flexible application systems, responsive to business needs, and capable of improving the productivity and effectiveness of an organisation's workforce.

The ASD framework is built on three levels, as shown in figure 9.6. The first is a set of guiding principles reflecting the needs of users and providing the basic foundation. These six guiding principles are:

- business orientation, to ensure that solutions address specific strategic and operational organisation objectives
- flexibility, for adapting to changing business environments, delivering high-quality solutions quickly and efficiently, and supporting executive decision making
- end-user participation, to ensure that solutions meet the day-to-day needs of business professionals

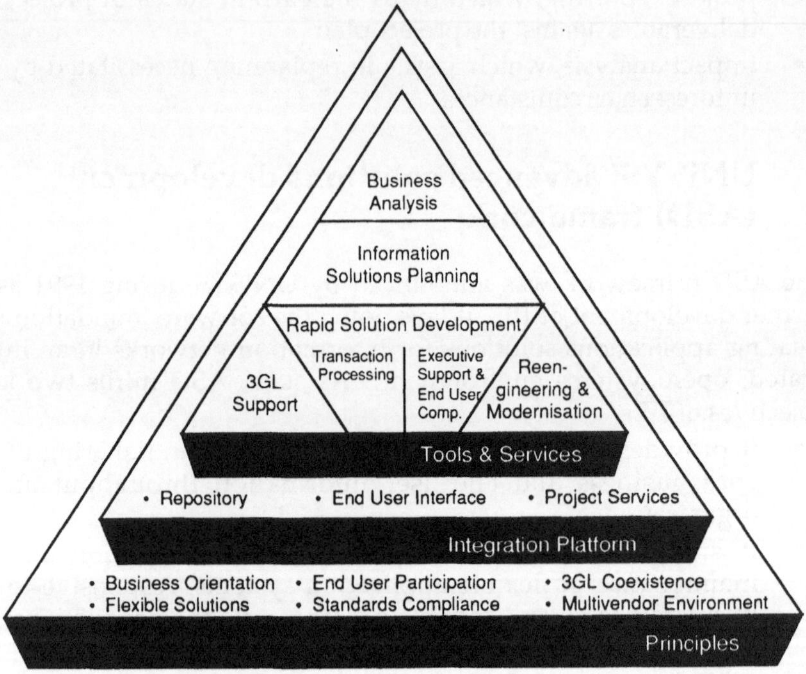

Figure 9.6 The ASD Framework

- standards compliance, to provide an open software environment that integrates UNiSYS, third party and user-built tools
- third generation language coexistence, to protect an organisation's investments in existing applications; and
- support of multi-vendor environments, to let users select and implement solutions on UNiSYS premium platforms as well as on non-UNiSYS platforms.

The second level in the ASD framework is the software integration platform, which provides a cohesive, integrated environment for developing solutions. Three major facilities tie all the tools together:

- the UNiSYS repository
- the end-user interface
- project services.

The UNiSYS repository is an organisation-wide system that stores, maintains, controls, protects, and shares solution development information. The object-oriented repository will be interoperable with other products and will support the ANSI/SPARC three schema architecture, which runs on mainframes, server platforms and workstations. At present, a workstation-based version of the repository is available as part of the Designer Workbench. The full three-tier version is expected by 1994.

The end-user interface is a single common graphical window to the tools and solutions, in both development and production environments. UNiSYS will support industry standard GUIs as well as multimedia. The current capabilities are included in Designer Workbench. This provides an integrated GUI with local graphical design capabilities, complying to industry standards and integrating LINC, MAPPER and ALLY applications in a co-operative processing environment.

Project services is a set of common management facilities, accessible to all tools in the framework. These include project management, security control, document definition and production, and workflow management. A variety of features and capabilities exist today in the information hub and workstation repositories, which provide support for such services.

Tools and services comprise the third level in the ASD framework. This includes all the tools and services needed for planning and implementing solutions, such as those for business analysis, information systems planning and rapid application development (RAD). These tools include not only front end and back end CASE products, but also executive support and end-user development tools.

The ASD framework is based on proven tools, including products such as LINC, MAPPER and ALLY. Several new products are also included, for instance Designer Workbench, which integrates new and existing applications on diverse platforms by implementing standard graphical user interfaces, such as Windows 3.

An important element of the framework is the involvement of third party companies, working in partnership with UNiSYS to provide a variety of key tools and services. These third party agreements focus on four areas:

- provision of business analysis and information systems planning services
- ensuring interoperability with other leading CASE tools
- joint development of advanced tools
- porting UNiSYS tools to other UNIX platforms.

In the UK, UNiSYS has been working with Coopers Deloitte from 1990 to 1992 to build business solutions based on Coopers' Summit methodologies (Summit S for strategic information systems planning, and Summit D for systems development and delivery). Automated support for these two approaches has been jointly developed. The Business Planning Workbench was released in March 1992. The Systems Design Workbench is still under development at the time of writing.

Other strategic partnerships have been negotiated by UNiSYS. There is a business alliance with KPMG Peat Marwick, which will initially focus on using LINC, MAPPER and ALLY to develop applications for the Fortune 1000 marketplace in the USA.

Also, an agreement has been reached for software development and licensing with Cadware, who will develop front-end CASE products that will integrate with LINC, using Systems Development II; Cadware will also adapt Systems Developer II for other UNiSYS environments and UNiSYS will license it for internal software development

There is also an agreement with Procase Corporation, covering the development of a COBOL version of Procase's SMARTsystem software, maintenance and reverse engineering tools to be marketed worldwide by UNiSYS, and co-operative technical and marketing

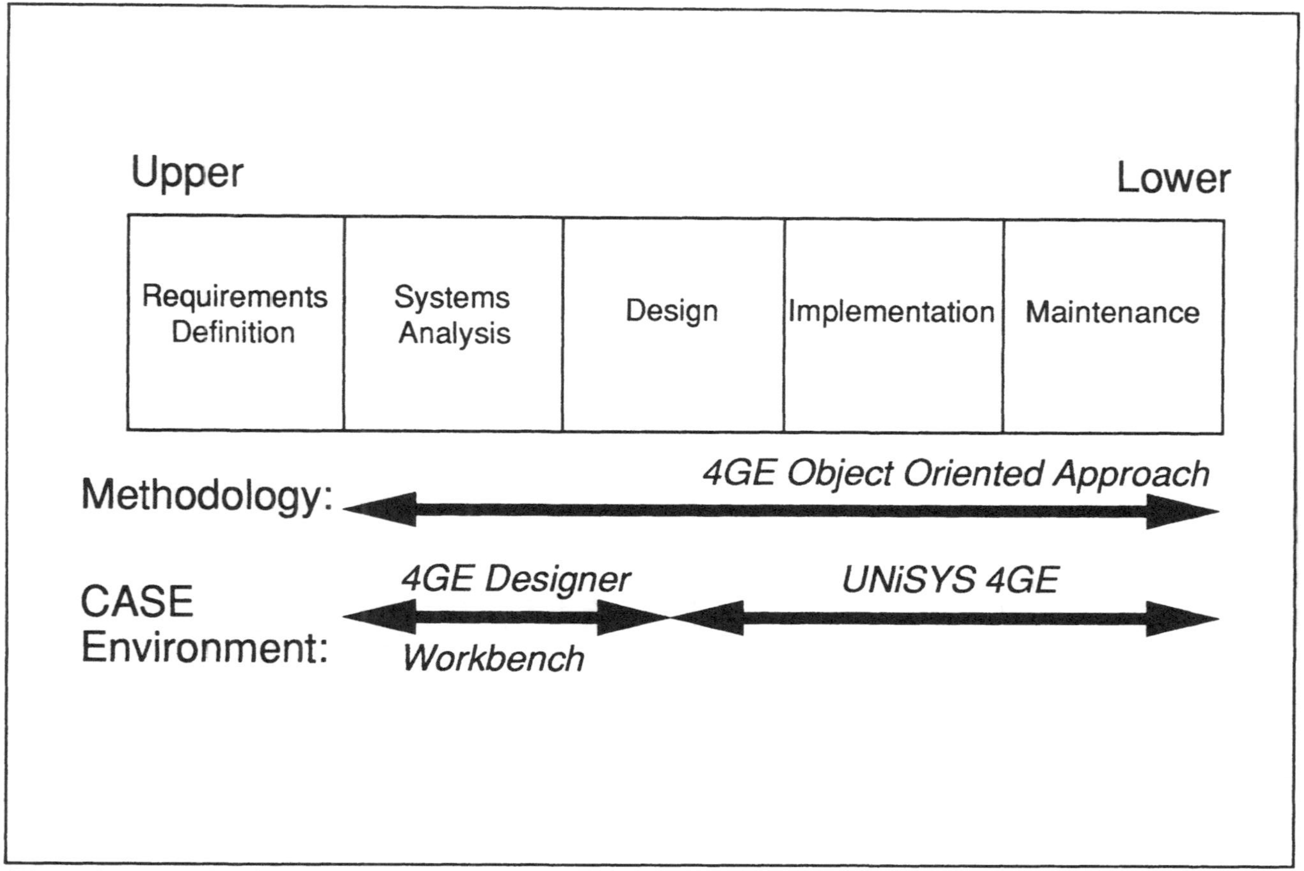

Figure 9.7 *UNiSYS' Proprietary Direction for I-CASE*

agreements have been set up with Intersolv, Knowledgeware, LBMS and Gamma International.

The platform provided by the ASD framework can be viewed as an open IPSE, able to accommodate other technologies such as knowledge-based systems, imaging and office automation.

5 UNiSYS' strategic direction for CASE

The UNiSYS strategic direction for CASE has been driven by the following factors:

- different methodologies dominate different market places
- the key front end CASE tools vary from country to country
- users need to generate total systems, not just programs
- users require applications that are portable across different hardware platforms
- users wish to be able to choose the most appropriate CASE products to help them
- users require a homogenous development environment, covering the whole development life cycle.

When UNiSYS announced its CASE strategy, on October 30th 1989, it also advertised extensions and additions to the LINC and MAPPER CASE/4GL product sets. Commitment was made to implement an open UNIX and CASE/4GL environment, and UNiSYS announced a series of initiatives with European CASE suppliers, including those suppliers interfacing their tools with LINC.

Hugo Simpson, UNiSYS' Vice President for Information and Productivity Systems, noted that 'while developers need diagrams and pictures to facilitate the analysis and design phases of the application development process ultimately they must translate those objects into code. An application is ultimately code ... not pictures.'

UNiSYS has expanded the capability of its fourth generation environments, MAPPER and LINC, to support the entire life cycle of an application, and now offers two integrated CASE packages based on these products. This required the development of two upper CASE products – Front End MAPPER and LINC – which also support specially developed, 4GE-oriented approaches which maximise the capabilities of MAPPER and LINC. Figure 9.7 illustrates the implementation components.

The implementation of the open CASE interface will allow users to interface any front end CASE tool to the UNiSYS 4GE. Figure 9.8 illustrates the architecture of the open CASE interface and its components, which are:

- analyst/designer workbench
- 4GE open CASE interface
- loader
- fourth generation environment (4GE).

Using the analyst/designer workbench, the developer analyses and designs the required application, which results in the definition and description of the programs, modules, screens, databases etc. This information is captured in the upper CASE data dictionary (DD) and must be transferred to the lower CASE environment.

The 4GE open CASE interface utilities transfer the information from the upper CASE repository, and supplement it in a manner that conforms to the concepts and rules of the 4GE used as the lower CASE implementation vehicle. In the next step, the information is transformed to the format expected by the repository of the 4GE. These utilities are developed jointly by UNiSYS and the upper CASE vendor(s).

The loader utility loads the information into the data dictionary or repository of the 4GE. This utility is developed by UNiSYS and is provided as part of the 4GE offering.

In the fourth generation environment (4GE), the developer completes the definitions of the upper CASE product using the 4GE facilities. The 4GE then generates the complete system, containing programs, database, screens, reports, documentation and so on.

6 Summary of UNiSYS' architecture and CASE strategy

IIE and the ASD framework together represent UNiSYS' first unified architecture; this embodies the company's strategy to turn around its fortunes. The IIE strategy recognises the fact that UNiSYS does not set agendas in corporate information processing. For users trying to leverage an installed base of mainframes, IIE makes a lot of sense, and for their existing customers it may represent a salvation.

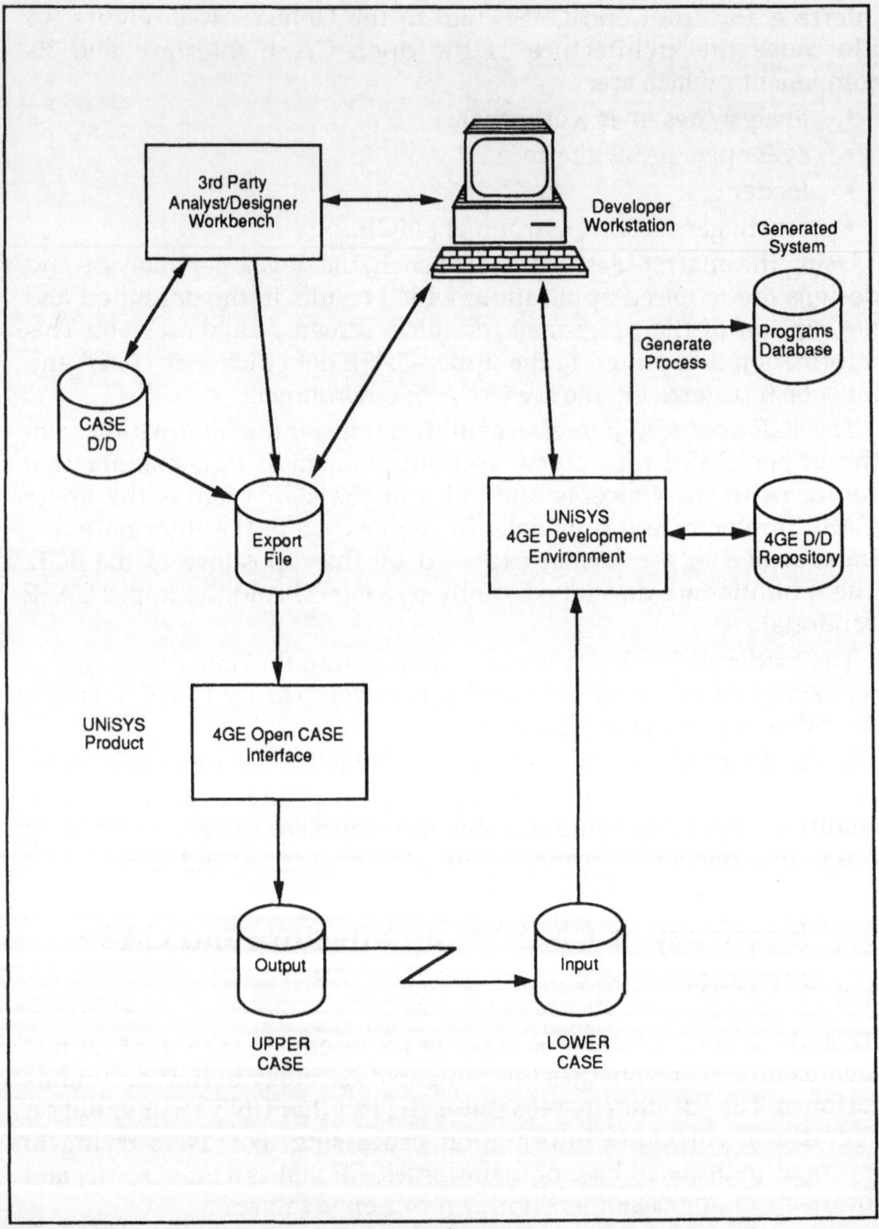

Figure 9.8 UNiSYS Open CASE Architecture

UNiSYS has tried to differentiate itself by basing the ASD framework on a new business-oriented approach to systems development, rather than on a conventional systems development life cycle. This involves ongoing, and very active, end-user participation. UNiSYS sees a large hole in the current CASE market, in that tools do not support end-user computing. The new technologies that UNiSYS is introducing will enable the use of new methods and promote improvements in application quality and productivity. These technologies are:

- object orientation, as a means of defining data and business rules and procedures
- artificial intelligence and expert systems, as a means of supporting all other technologies
- ICONIC, to provide a transition from diagrammatic design to visual programming
- a single repository, as the storehouse for information, and as the linchpin for integrating software development tools
- reusability, as a means of preventing people re-inventing the wheel
- re-engineering, to preserve the investment in code whilst taking advantage of new technologies.

Problem areas remain, however, when it comes to interfacing the 'code it and see' philosophy of the LINC Design Method with the structured analysis and design approach of front end CASE tools: they do not sit easily together. Whether UNiSYS can achieve a resolution remains to be seen.

7 References

1. UNiSYS (1990), *Integrated Information Environment, Technical Overview*, 4133 1497-000,9/90
2. Network Monitor (1990), *UNiSYS IIE*, Patricia Seybold's Network Monitor, Vol 5, No.10, October 1990
3. James Martin Associates (undated), *ASD Framework*, UNiSYS 4124 4104-000
4. Owen, Kenneth (1992), *Moving from strategy to reality*, UNiSYS house magazine, 1st quarter, 1992

Chapter 10

Hewlett Packard: NewWave and CASEdge

1 Introduction

Hewlett Packard (HP) has a distinctive approach to CASE, which can be summarised by the following intentions:

- to offer complete solutions, including consultancy help with planning, education and support
- to build upon their own experience and knowledge of methodologies, languages, object-oriented programming and software metrics
- to work towards open-industry standards.

HP's CASE strategy has been given the name CASEdge. It consists of three elements:

- CASEdge/Solutions (also referred to as Environments in some HP literature) – a series of software engineering environments, available from HP and independent software vendors (ISVs in HP terminology), that provide solutions for the specialised needs of organisations; HP's open systems environment is based on SoftBench
- CASEdge/Tools and Languages – libraries of specific tools developed by HP and third party vendors that may be selected and integrated under SoftBench to create a customised CASE environment

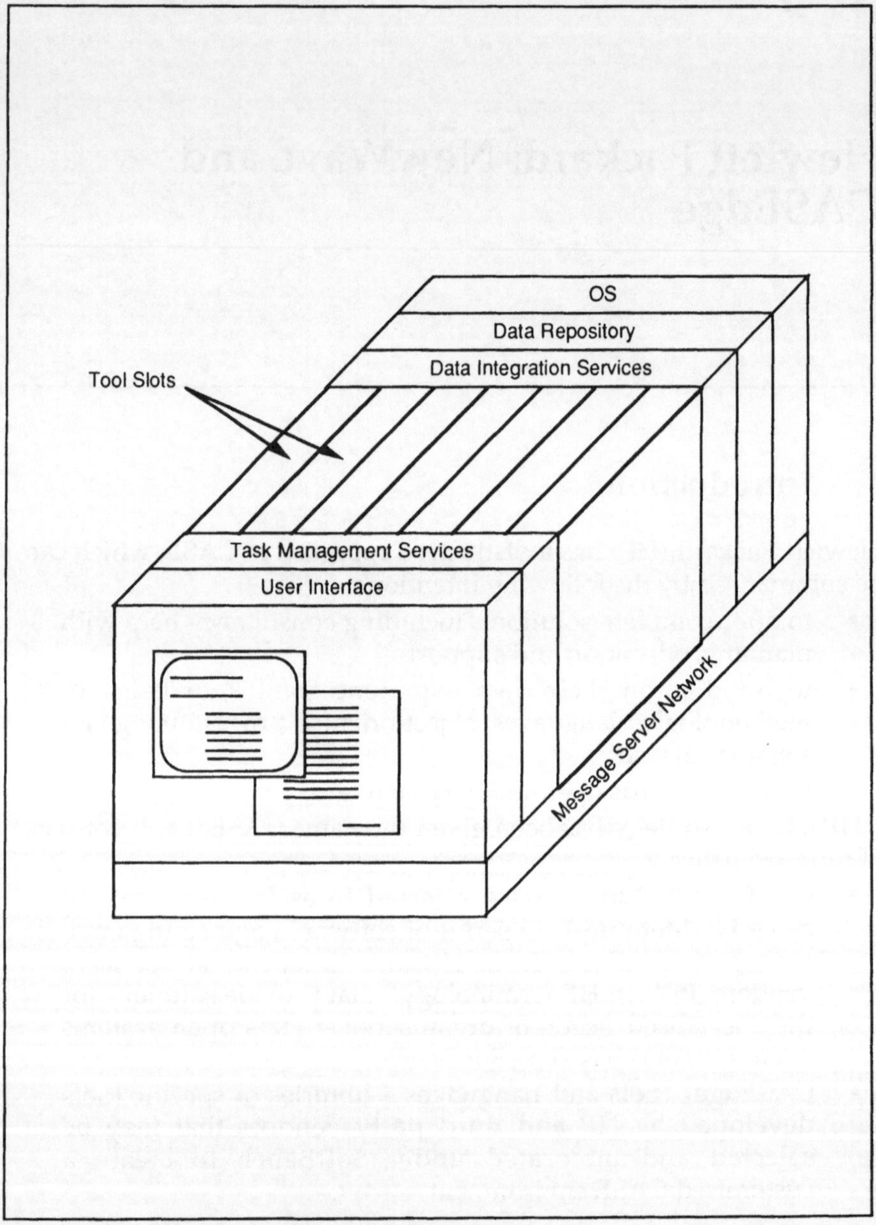

Figure 10.1 The HP CASE Environment Model

- CASEdge/Knowledge – a set of software engineering consulting services covering assessment, metrics, education, methods and implementation assistance.

2 CASEdge/Solutions: SoftBench, Encapsulator and the CASE environment model

SoftBench is based on an automated design technology model called the CASE environment model (see figure 10.1). This presents an framework for coordinating and controlling the behaviour of a set of cooperating tools or applications within the integrated environment. These tools can be selected and integrated according to user needs.

The model is based on the assumption that most implementations will be physically distributed, so that individual tools in a set can run on any number of machines on a network. Support for multiple data repositories is an important attribute of the CASE environment model.

The model has been adopted by a number of standards groups as a reference model for evaluating different options prior to agreement on a standard. These include:

- the CAD Framework Initiative (CFI)
- the European Computer Manufacturers Association (ECMA)
- the National Institute for Standards and Technology (NIST)
- IWCASE Inc (International Workshop on CASE)
- the Japanese SIGMA project.

SoftBench provides five tools which target the construction, test and maintenance phases of the systems development life cycle. These are:

- program editor
- program builder
- program debugger
- static analyzer
- development manager.

The program editor, a language sensitive mouse/menu-based source file editor, allows the user to customise PF key usage. Additional editors can be integrated into the environment, which supports synchronised file view updates.

The program builder automates the process of compiling programs composed of many different source files. Only where source files have been modified since the last compile are they recompiled. It presents compilation errors in a browser format, allowing a separate window to be brought up where source code may be edited. SoftBench allows a system to be configured in a network as a centralised compilation server, thus off-loading intensive CPU usage from a development workstation.

The program debugger is a multi-window tool, designed to help the understanding of a program's dynamic execution. Windows are provided for showing source code, information on tracing and program I/O. Within source code views, a pointer is provided that indicates the next line to be executed, and breakpoints can be interactively set by pointing at the line with the mouse. The debugger can display dynamic changes in variable values, function and procedure calls, and changes to the subroutine stack during program execution. Dynamic changes in the CPU and FPA registers can be displayed graphically, and machine-level instructions traced. The steps taken to achieve a given execution can be recorded, and the state recreated.

The static analyzer provides information about the structure of a program. It satisfies cross-reference queries such as 'where declared', 'where defined', 'where used' and 'where modified'. Function queries also are supported, and these include 'what functions are defined' and 'what calls are made in this function'. The analyzer, in addition, handles general pattern search queries.

The development manager acts as 'home base' from which all the other tools in the environment can be executed. It manages the versions of files on which other SoftBench tools operate. The version control facility provides multiple revisions, access control and branch merge support. File check-in and check-out can be accessed from other SoftBench tools.

The HP SoftBench tools are integrated via the tool integration platform, a framework that provides SoftBench with several additional features. It allows for the integration of tools and is responsible for providing:

- distributed computing services
- tool communication
- OSF/Motif appearance and behaviour across all tools
- integrated on-line help.

The SoftBench environment supports distributed execution, distributed data and distributed display. SoftBench tools can execute on any host in the network, provided that the host has SoftBench installed, and remote execution is transparent. A SoftBench tool can access the desired data no matter where the tool is running. This is enabled through the use of either the network file system (NFS) or HP's remote file access (RFA). SoftBench is built upon the X-Window system, which allows programs to execute on one system and display and support user input/output on another.

SoftBench tools communicate in a networked, heterogeneous environment via a broadcast communication facility that supports close communication between independent tools. Tools provide compatible user and program access to functionality. Message requests allow one tool to invoke the functionality of another tool, and notification messages allow the tool or the user to define the necessary triggers, which respond to events and initiate actions. Triggers are cause-and-effect relationships, and are used to link one or more tools together to support a task or process. Users may bring their own tools into the SoftBench environment by using HP Encapsulator.

The integrated on-line help facility cooperates with other tools in the environment to service requests for help. Any of the SoftBench tools can obtain general and context-sensitive information, and definitions of terms used. Figure 10.2 portrays the SoftBench environment.

HP Encapsulator allows users to extend and customise the SoftBench environment. Encapsulator can add user developed tools, without source code modification. using Encapsulator, the environment can be customised to automate the development process by establishing communication links between SoftBench and the user-added tools.

Encapsulator provides a C-like specification language, called encapsulator description language (EDL), which allows the user to define an EDL file. This file defines the window-based interface to an application, the program's message-based interface, the triggers and the actions which control the encapsulated tool/program. The EDL file is compiled with Encapsulator.

Using tool encapsulation, non screen-oriented products, developed in the standard UNIX pipe model, can use the SoftBench tool integration platform facilities. There are six steps to encapsulating a tool:

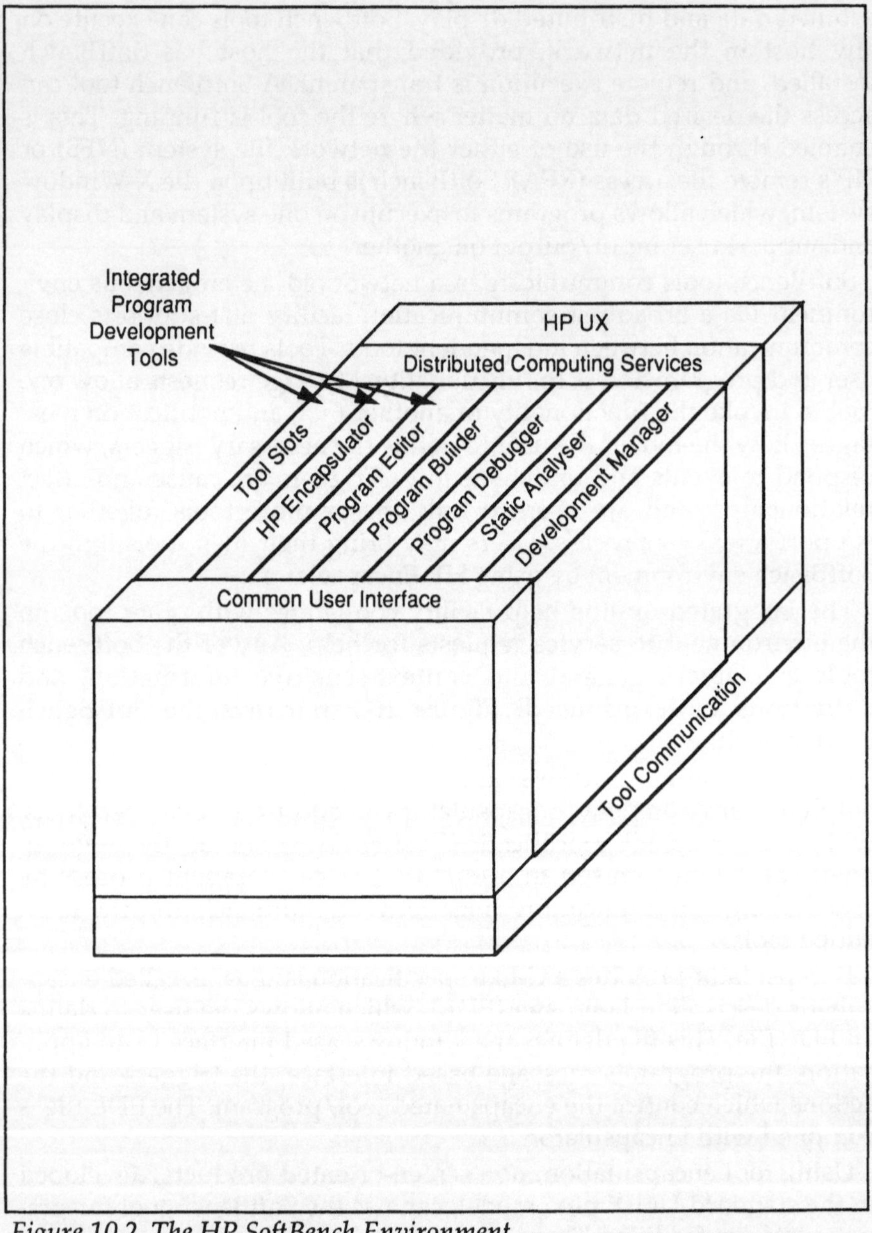

Figure 10.2 The HP SoftBench Environment

- determine if the tool fits the standard UNIX pipe model
- understand the functionality of the tool, especially the user interaction
- use EDL to define:
  - the actions the tool will perform
  - the user interface
  - the message interface
  - the triggers.

SoftBench and Encapsulator were formerly restricted to the HP-UX platform. However, HP ported them to run on the Sun SPARCstation platform from January 1991.

3 NewWave computing

NewWave is, in effect, HP's computing strategy. It is an application environment that provides a base set of facilities to allow users to focus on results rather than on learning and using individual applications. It offers an open, integrated and consistent application interface based on industry standards for software and hardware. The interface technology in NewWave allows users to manipulate complex objects such as documents, rather than programs and data files.

The first thing that a user sees is the NewWave office window. It is modelled on a real office and contains a file drawer, waste basket and other tools. It is the starting point of all tasks, and users can create, file, copy, delete and rename objects from this window.

The object management facility (OMF) is one of the key technologies in NewWave. It links data to applications to create an object. An object is data bound together with the application necessary to manipulate the data, and can be represented by an icon. The OMF allows a user to show the same information in multiple places on the PC.

NewWave has the capability of setting up hot links, whereby information is shared between different objects. When the information is updated, all the objects using that information will automatically have access to the latest version.

NewWave has an agent tasking capability, where the user can train their agents to automate tasks across applications by simply going through the task once. This produces an agent task, in effect a macro.

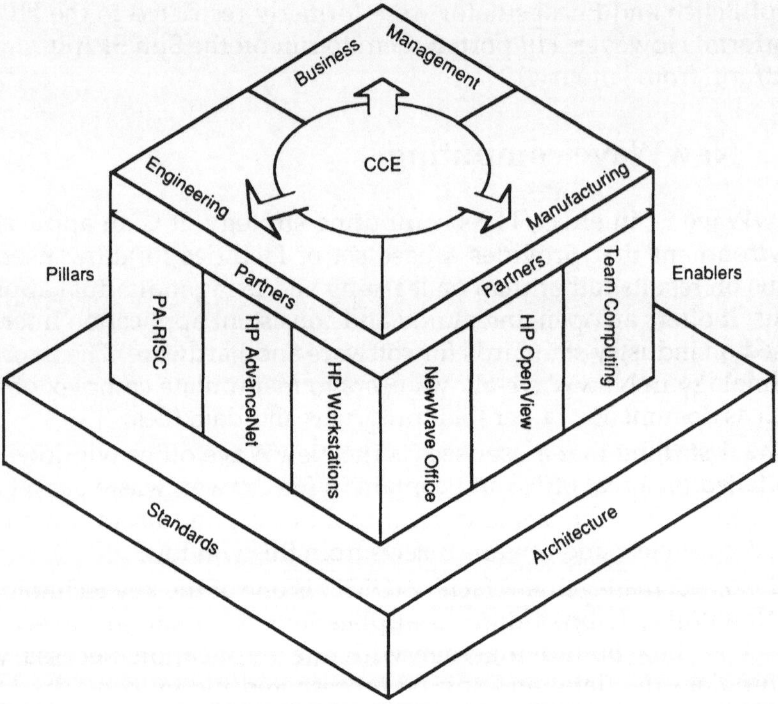

Figure 10.3 HP's NewWave Architecture

An MS-Windows/MS-DOS version of the NewWave developers kit has been available since 1988. This includes the necessary software and supporting manuals to enable a developer to write applications that take advantage of NewWave facilities. Through an encapsulation process, NewWave users can use MS-DOS and MS-Windows software. Two levels of encapsulation are available.

- The DOS programs menu provides the user with access to their MS-DOS or MS-Windows applications through a list of application titles. These appear without icons in a DOS programs menu available from the NewWave office.
- Icon-level encapsulation allows the user to create new data files, which then appear as icons in NewWave office. A bridge builder object allows users with a knowledge of MS-DOS and MS-Windows to create their own icon-level encapsulations.

Figure 10.3 shows the architectural picture for NewWave.

NewWave fits into the CASEdge strategy by providing operating system and data integration services, as well as task management and the user interface parts of the CASE environment model.

4 Review of HP strategies

HP has concentrated on exploiting a PCTE+ type environment to specify the interface between the software engineering toolset and the underlying operating system. The granularity of the objects addressed by the architectures is at a higher level than for most other vendors. Despite this, HP succeeds in offering users the type of environment advocated by the authors in 1991.[1]

The problem lies, at present, with the other tools that fit into this environment: they are predominantly real-time systems based, and thus unfamiliar to most in the commercial systems world.

In addition, there is little assistance for the early stages of the life cycle (particularly strategic planning and analysis), another serious constraint on CASEdge's wide adoption.

5 References

1 Holloway, S. and Bidgood, T. (1991) *CASE Handbook for Information Managers*, Avebury Technical
2 HP (1989), *CASE Primer*, HP CASEdge, 5953-6395
3 HP (1989), *Software Engineering Productivity*, HP CASEdge/Knowledge, 5952-0225
4 HP (1989), *Software Development Catalog*, HP CASEdge/Tools, 5952-0221
5 HP (1989), *HP SoftBench 1.0 for the HP9000 Series 300/800: Technical Data*, 5952-0222
6 HP (1989), *HP Encapsulator for HP9000 Series 300/800: Technical Data*, 5952-0223
7 HP (1990), *Systems News*, Issue 2,NSD09, April 1990
8 HP (1990), *NewWave Computing: Technical Supplement*, NSD09 supp., April 1990
9 HP (1990), *HP NewWave: Technical Data*, 5952-1565
10 HP (undated) *CASEdge: The only totally open, integrated approach to CASE*, (video)

Chapter 11

Bull: Distributed Development Environment (DDE)

1 Introduction

The distributed computing model (DCM) is Groupe Bull's response to challenge of delivering enterprise computing resources to the user desktop. To understand their requirements, Groupe Bull talked to its customers, as well as to third party software vendors.

The message received was that users want, fundamentally, to deliver business solution applications to the desktop. Software developers want to focus development on business problems rather than re-inventing common functions, and therefore they need building blocks for information systems. The third party software vendors required open access to application programming interfaces (APIs) at all levels.

The DCM provides the specifications of an open and logically distributed systems architecture to guide the evolution of Bull products. The model is open because it is based on a set of *de jure* and *de facto* standards, so that systems from Bull can work effectively with products from other suppliers. It is logically distributed, offering different degrees of centralisation and decentralisation of data and processing, depending upon organisational choice for the deployment of information systems.

DCM takes into account three user perspectives, each of which has its own specific requirements, but which all need to use the same integrated systems as a working tool. For the end user, the model means information on the desktop, with all the applications and information resources available. For the system administrator, it means the management and security of the information systems considered as a whole. For the application developer, it means services for building applications.

The model provides modularity so that a coherent set of architectural building blocks are available. The support for open systems means that interfaces to each service are supplied and documented. These interfaces are based on de jure or de facto standards, so as to allow interoperability, porting and integration of software. End users have transparent access to applications on multiple platforms, and applications may use services distributed throughout the network.

2 The DCM architecture

The distributed computing model is represented as a set of components, each of which describes a family of elements (often services) with a common functional role. It consists of four horizontal components:

- the applications component, representing the applications available to end users to meet business needs. and containing both generic and industry-specific applications
- application services, enabling applications by the capture, restitution, sharing or processing of information
- distributed services, providing transparent distribution of functions, and in particular application services, across the network
- communications and system services, providing the actual transport of information and the system processing power.

These components are covered on either side by two vertical components:

- the integrated systems management and security component, providing the applications and the common services for security and the management of the distributed environment
- the application development component, including a variety of services and tools (from software partners and/or Bull) required for a complete application development environment.

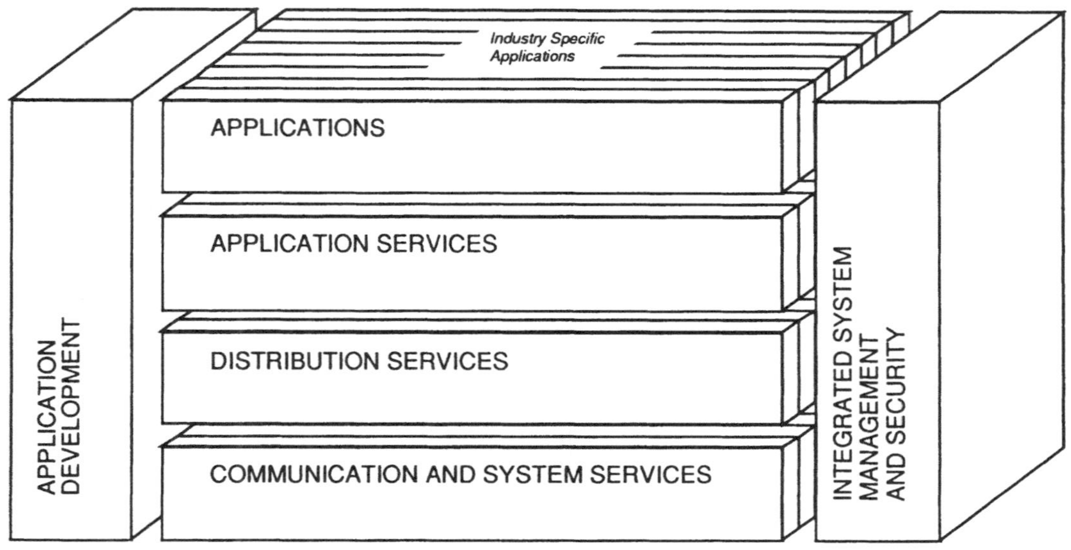

Figure 11.1 Bull's Distributed Computing Model

The model is shown in three dimensions in order to represent – as complete vertical planes – its specific implementations for particular industry types, such as manufacturing, banking, insurance and distribution. Figure 11.1 shows the model diagrammatically.

3 DCM topology

Bull products are to evolve within the topology depicted in figure 11.2. The product offered is based on 'the progressive logical unification of the distributed systems which will share an increasing number of application services'.

The servers and workstations on an organisation's information network are distributed geographically, and the different systems are connected through a LAN and/or a WAN. Any protocol differences are masked by the communication services, which present common interfaces to the distribution services.

The distribution services provide unified naming and file systems, consistent time, and function location transparency for the application services.

The application services can make common functions available to applications in a transparent manner. Each application service is generally located on the server which best suits its requirements for processing power, communication capabilities, I/O speed, etc.

The end user has access to the organisation's information systems through a graphical user interface on PCs, X-Window terminals or workstations, or through a VDU character-based interface.

The coherence of all the components within a distributed systems domain is ensured by the integrated systems management, which provides a single, logical view of the domain as a whole. Security is ensured through a common security administration for the centre domain.

4 The applications component

The applications comprised in this component are built upon available interfaces. They call the necessary underlying services and interact with the end user through the presentation service. Broadly,

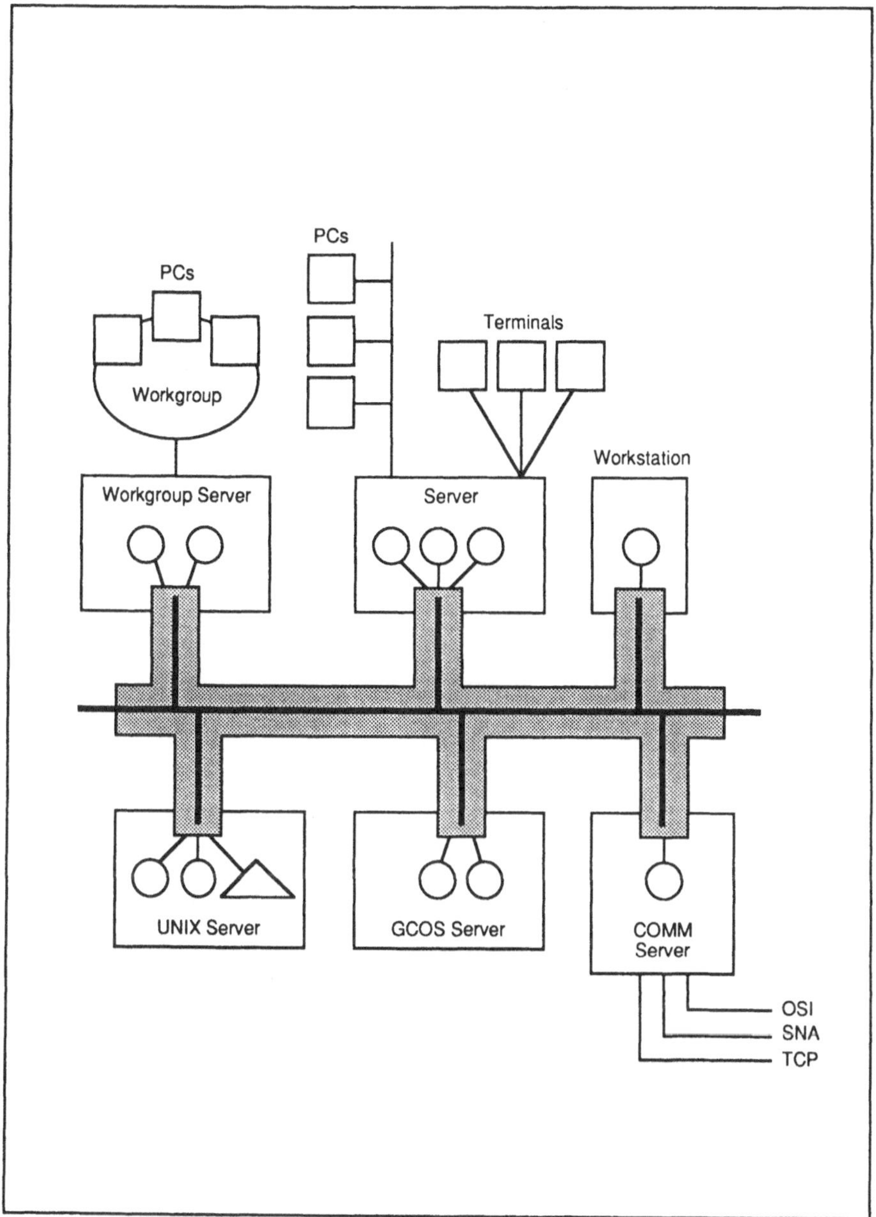

Figure 11.2 Bull's Distributed Computing Model Topology

the applications offered by Bull themselves can be split into two classes:

- generic applications, such as office automation applications, providing general business functions and applicable across many types of businesses
- industry-specific applications, used in one particular sector of industry.

5 The application services component

The application services component consists of services which enable applications by the capture, restitution, sharing or processing of information. The functions offered by the services are presented in a transparent way, and free the application developer from having to know anything about the physical location or particular details of each service. They are software elements which can be shared by all systems, which need not be replicated in applications. The physical location of the application services is determined by the systems administrator.

END-USER SERVICES	EXCHANGE SERVICES	LIVE DATA SERVICES
PRESENTATION DIRECTORY	MAIL	DOCUMENT FILING/RETRIEVAL
PRINT	EDI	DATABASE ACCESS
SCANNING	FILE TRANSFER	TRANSACTION PROCESSING
	DOCUMENT INTERCHANGE	WORKFLOW

Figure 11.3 Generic Application Services

Some applications services, such as the presentation service, perform tasks which benefit individuals, or one application program at a time, providing a specific representation of data. Other application services, such as the mail service, benefit a group of individuals or applications. A final set contains services such as the database service, performing further processing on the organisation's pool of data, and therefore benefiting the whole organisation.

As currently defined, the available generic application services are shown in figure 11.3.

End user services satisfy a user's or application's need to view, represent, print or capture information. They operate on the human computer interface. The presentation service is supported by a number of APIs to allow the use of standard GUI products. It includes a suite of tools and methods, including a style guide. Figure 11.4 shows these interfaces.

PLATFORM	API	SUPPLIER
MS-DOS	Windows 3	Microsoft
UNIX workstation & X-Terminal	X Window MOTIF 1.1	MIT OSF
OS/2	Presentation Manager	Microsoft
All	Distributed Dialog Manager	(New industry standard)

Figure 11.4 Presentation Service APIs

Exchange services ensure the necessary exchange of information for interactive cooperation between users, workgroups and applications. They operate on organisational and information system boundaries, and consist of:

- the directory service, offering full X.500 functionality
- the mail service, based on the X.400 recommendations for message handling
- the EDI service, supporting standards for several formats:
  - EDIFACT
  - ANSI X.12, and
  - TRADACOMMS

- the file transfer service, an implementation of the OSI FTAM standard, and
- the document interchange service, which specifies ISO ODA/ODIF as the preferred format.

The live data services operate on the permanent information of an organisation. These services are 'mission critical' applications in a distributed information system. The document filing/retrieval service has two classes:

- folder management, based on the ISO DFR model, to manipulate a hierarchy of designated objects such as filing cabinets, folders and documents, and
- content management, dealing with the attributes of documents and their internal structures.

The database service provides access to all types of database, be they private, departmental, corporate or public. Whilst support for existing dbms products is there, the model specifies a single preferred SQL interface, which is mandatory for applications requiring distributed and transparent access to heterogeneous databases. This preferred interface conforms to current and emerging standards for access to relational databases (SQL Access Group and X/OPEN).

6 The distributed services component

The distributed services component provides for the transparent distribution of functions across the network, allowing access to a wide variety of services and applications, without the user needing to worry about the location of the service or the peculiarities of the platform. These services are the backbone of the distributed computing environment, allowing applications and application services to be distributed, based on a conventional client/server model. The services include:

- distributed file
- naming
- remote procedure call (RPC)
- time.

The distributed file service is inherent to the distributed file system (DFS) implemented in the DCM, and uses OSF's DCE technology to provide crash recovery, a physical file system and a protocol exporter.

The naming service maps user-oriented names or objects of interest in a distributed computing environment into computer-oriented entries in a distributed database. The objects to be named include countries, organisations, persons, groups, organisational roles, computers, printers, files, processes and application services. The components of the naming service are based on OSF's DCE technology and comprise:

- a local naming service
- a global X.500 service
- XDS and XOM (X/OPEN compliant) APIs.

The remote procedure call (RPC) allows programs running locally to call procedures implemented on remote systems. It is this mechanism which allows access to application services to be distributed, and it is used to export the API of the application service onto other platforms. The RPC service is based, once again, on OSF's DCE RPC, and includes two major components:

- a RPC facility, containing an automatic data conversion that masks the differences between data representations
- a compiler to convert high-level interface descriptions of remote procedures into portable C code.

The time service regulates the system clocks throughout the network, so that they closely match each other and provide accurate timings for all distributed processing.

7 The communications and system services component

The communications and system services component contains the services which provide the actual transport of information and processing power in a heterogeneous environment. The DCM supports a wide range of interconnection protocols, including OSI, TCP, and SNA.

DCM contains three major types of basic communications access.

- XTI (X/OPEN Transport Interface) provides a transport interface covering both ISO and TCP transport protocols in connection oriented and connectionless mode.
- CPI-C (X/OPEN program-to-program communication) allows programs to operate a synchronous dialogue with

other programs in a peer-to-peer mode, and provides functions which open, read, send and close a dialogue. These functions are completed in the TP environment with a conformation level and two-phase commit mechanism.

- PC communication and terminal emulation allows PCs to be connected to servers or mainframes through various mechanisms, each of which has its own paradigm:
  - PCs to GCOS, using an interface called UVTI (unified virtual terminal interface), implemented on DOS and incorporating access to interactive applications running on GCOS systems
  - PCs to IBM systems, using an interface called HLLAPI to emulate a 3270 terminal
  - PCs to UNIX or OS/2 servers, using Microsoft Lan-Manager interfaces (called Mailshot and NamedPipe) to give access to all workgroup applications
  - NetBIOS-API, supported to provide backward compatibility.

System services provide the basic functions required by applications and other services to interact with the system platform. These functions include the following capabilities:

- memory allocation
- creation and management of processes
- management of real-time events.

The different application architectures supported by Bull have various sets of systems services, tailored for the class of applications best supported by each platform. The platforms are divided into two groups:

- open systems architectures, including DOS, OS/2 and UNIX
- GCOS.

8 The integrated systems management and security component

The integrated systems management and security component (see figures 11.5 and 11.6) provides the applications and the common services for security and for the management of the distributed environment. Management and security applications may be provided by Bull, by third party vendors or by the customer, and tools are available to support the following areas:

- network management
- system management
- spool management
- storage management.

Integrated systems management is implemented by applying standards-based technologies in three major areas

- Individual management applications are used to perform management tasks. They shield the user from the detailed steps necessary to perform these tasks, and ensure the integrity of management data.
- Common management services support the coherent management framework by making programming interfaces available for use by management applications. They include sieves, a script facility, an inference engine and event services. These services are presented through a program interface and use relevant standards, such as OSI CMIS/CMIP and Internet SNMP.
- Managed objects are a representation of resources within the computing environment.

The security services for operating systems adhere to draft standards which are currently being defined by the US DoD and the EEC. These standards comprise:

- access control
- audit
- integrity and data origin authentification for user requests and responses
- protection mechanisms for the denial service.

The security services operate within a security cell, which is a set of protected elements following a common security policy. These services comprise three main parts:

- secured objects
- common security services
- security applications.

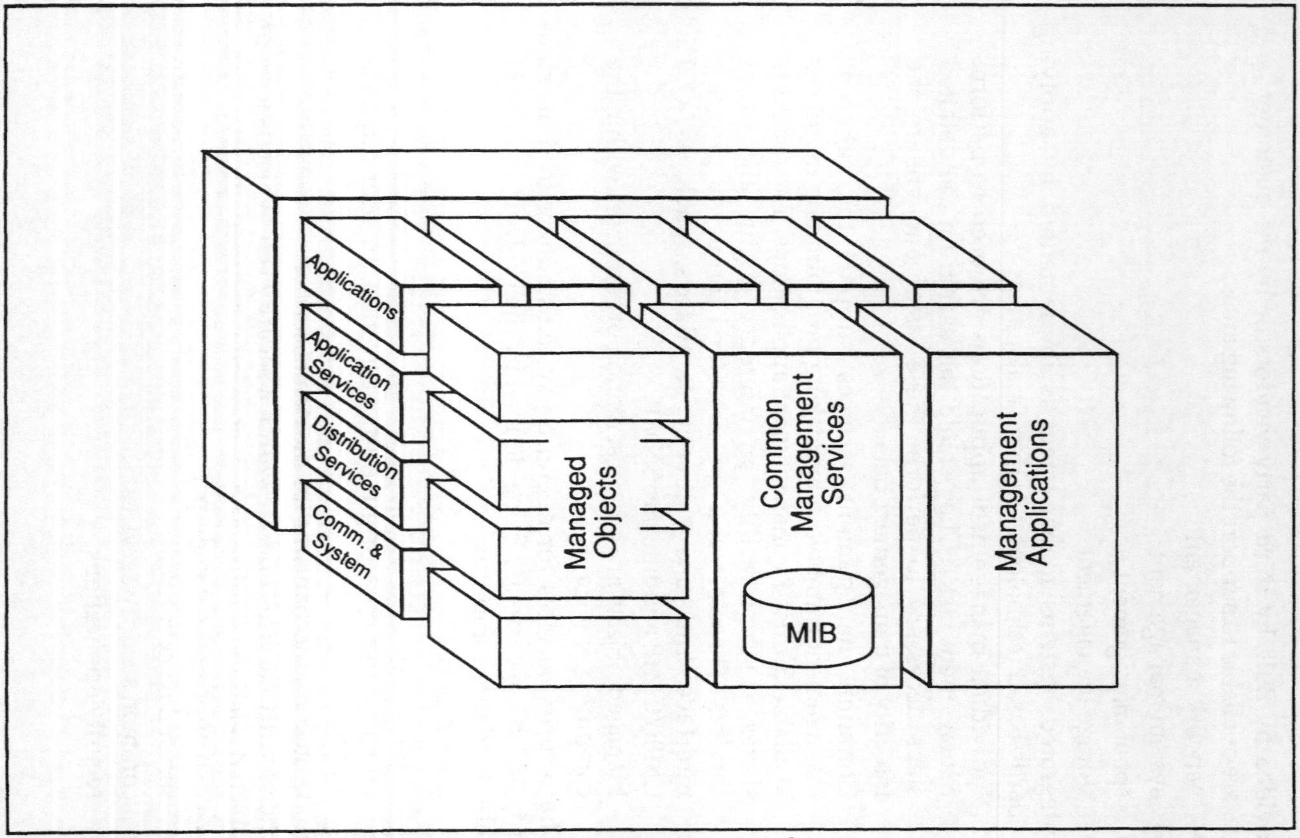

Figure 11.5 The DCM Integrated Systems Management Framework

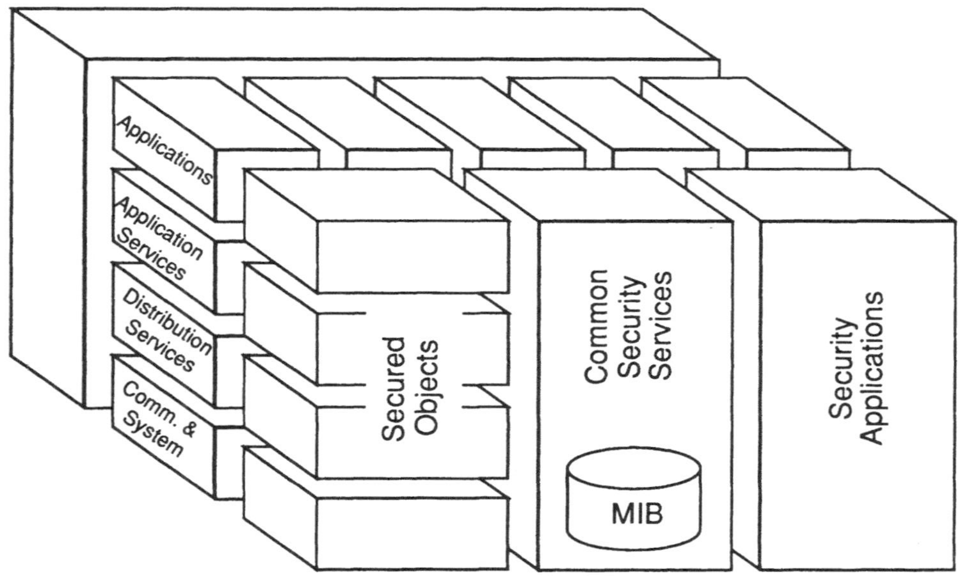

Figure 11.6 The DCM Security Framework

The model provides mechanisms to ensure the appropriate security for all software and hardware objects in an organisation's information systems. These secured objects include messages, databases, communication paths, servers and user applications.

The common security services handle the programming interfaces used by security applications, and include:

- identification
- authentification
- authorisation tools
- audit
- integrity
- confidentiality
- a security information base.

The availability of APIs to common security services enables customers to use the security framework to support their own security policy features. Bull security applications include:

- a secure welcome application
- specific applications using CP8 smart card capabilities
- secure user management
- a security supervisor, providing administration and control.

9 The application development component

The goal of the application development framework (see figure 11.7) is to 'industrialise the application development cycle, and to mask the complexity of the underlying operating system and distributed computing architecture'. The framework provides developers and third party vendors with:

- a set of standard programming languages and development tools
- access to all components of the distributed environment through open standard APIs
- an environment for integrated CASE, using a common repository and support for the full systems development life cycle in an IPSE.

The server and client operating environments for DCM are shown in figure 11.8.

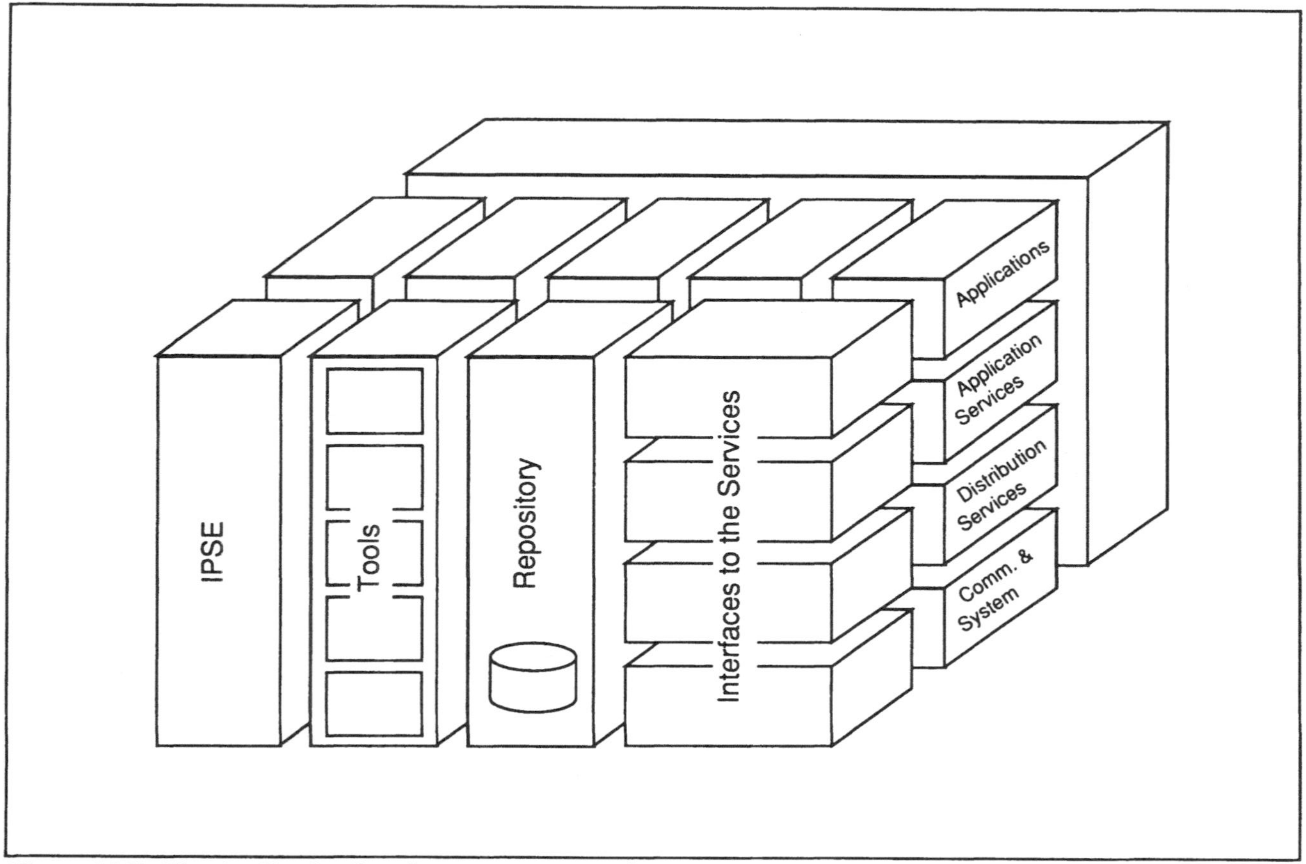

Figure 11.7 The Application Development Framework

SERVER OPERATING ENVIRONMENT	CLIENT OPERATING ENVIRONMENT	MAIN EXECUTION ENVIRONMENT
GCOS 8		TP8
GCOS 7		TDS
GCOS 6		HVS
BOS		X/OPEN POSIX
	MS-DOS	WINDOWS 3
	OS/2	PRESENTATION MANAGER

Figure 11.8 Server and Client Operating Environments for DCM

Bull's repository product will provide de facto standard interfaces including CPI-R, ANSI's IRDS and ECMA's PCTE.

Bull is to support, through the DCM, the following languages:

- ANSI C (X3.159- 1989)
- ANSI COBOL (X3.23- 1985)
- ANSI FORTRAN (X3.9- 1978)
- ISO PASCAL (7185- 1983).

10 References

1 Bull HN Information Systems Ltd (1991), *Distributed Computing Model: Solving Problems in the 1990s*, March 1991, YW24.00

2 Bull SA (1991), *Distributed Computing Model Specifications Overview*, March 1991, 23 A8 03 CS

Section C

Software Vendors' Strategies

Chapter 12

The Software Vendors

The choice of software vendors to be covered in this book was never going to be easy. By comparison, there is a limited number of significant hardware vendors – some people would identify only one, others four or five, or up to ten, 15 or 20. Whichever way you look at it this is a manageable number. The authors picked the half dozen most significant hardware vendors fairly easily.

For software vendors the total number of vendors exceeds 1000, though the feverish scramble of the late 1980s, which saw around 2000 vendors of back end tools alone, has started to rationalise: the long-promised 'shake out' of the industry is occurring with a vengeance, as new alliances, takeovers and bankruptcies are everyday news.

But which are the *key* vendors in the CASE software market? In the preface, we referred to the 'visionaries' and the others. Of the thousands of CASE software vendors, only a handful – more than the number of hardware vendors but still less than 25 – can be said to have any real vision of their own. Others, such as Knowledgeware or Bachman, are happy to fill a niche rôle in a hardware vendor's strategy (in IBM's AD/Cycle for both cases quoted).

At the time the book was conceived, certain vendors, such as Index Technology, had a glorious past but did not, in our view, have an obvious future. The formation of Intersolv, from the merger of Sage and Index, has partially rectified this situation, but we wait to see whether real success, to compare with past glories, will follow.

Other vendors, as diverse as Sapiens, Intellicorp and Cincom, have demonstrated that they have the vision. But they have not, arguably, achieved a degree of market penetration for their strategic products which can justify their inclusion here. We believe that they can succeed, as they deserve to on the grounds of innovation and initiative, and hope to include them in a future edition of this book.

Three of the vendors selected – CA, Software AG and MSP – have developed approaches (or visions) broadly similar to those of the hardware vendors, intended to build on a large and solid installed base. The other four, each with a vision and real credibility in the marketplace, are Oracle, Softlab, LBMS and Texas Instruments. Each of these offers a distinctive strategy, based around a single product or family of products.

These seven – featured in chapters 13 to 19 – are our *key* software vendors in 1992. We believe that at least four of them will even more significant in 1997 – unfortunately, we do not know which four!

Chapter 13

Computer Associates: CA90s

1 Introduction

Computer Associates has been the prime example of a vendor who acquires technology to gain market share. In 1987, it purchased UCCEL, followed in 1988 by Applied Data Research (ADR) and, in 1989, Cullinet. The latter two acquisitions have made CA a major player in the DBMS/4GL market place. The Cullinet take-over has allowed CA to diverge into hardware platforms other than IBM. However, this has at the same time put back its ACE strategy, devised after the take-over of ADR. This is because CA now has two DBMSs, two 4GLs and two dictionaries. 1990 saw CA announce CA90s, which further enhanced ACE, but did nothing to resolve the double product strategy. The subsequent acquisition of Pansophic, with its variety of products, has further muddied the waters. Geoffrey Finlay of Intersolv commented that 'CA is becoming like the Woolworth's of software.'[1]

2 CA90s: Computing Architecture for the 90s

CA90s is a software development architecture that represents a structured design and the principles of that design. The architecture is layered, as shown in figure 13.1, to allow the sharing of software services among applications and to provide the ability to process

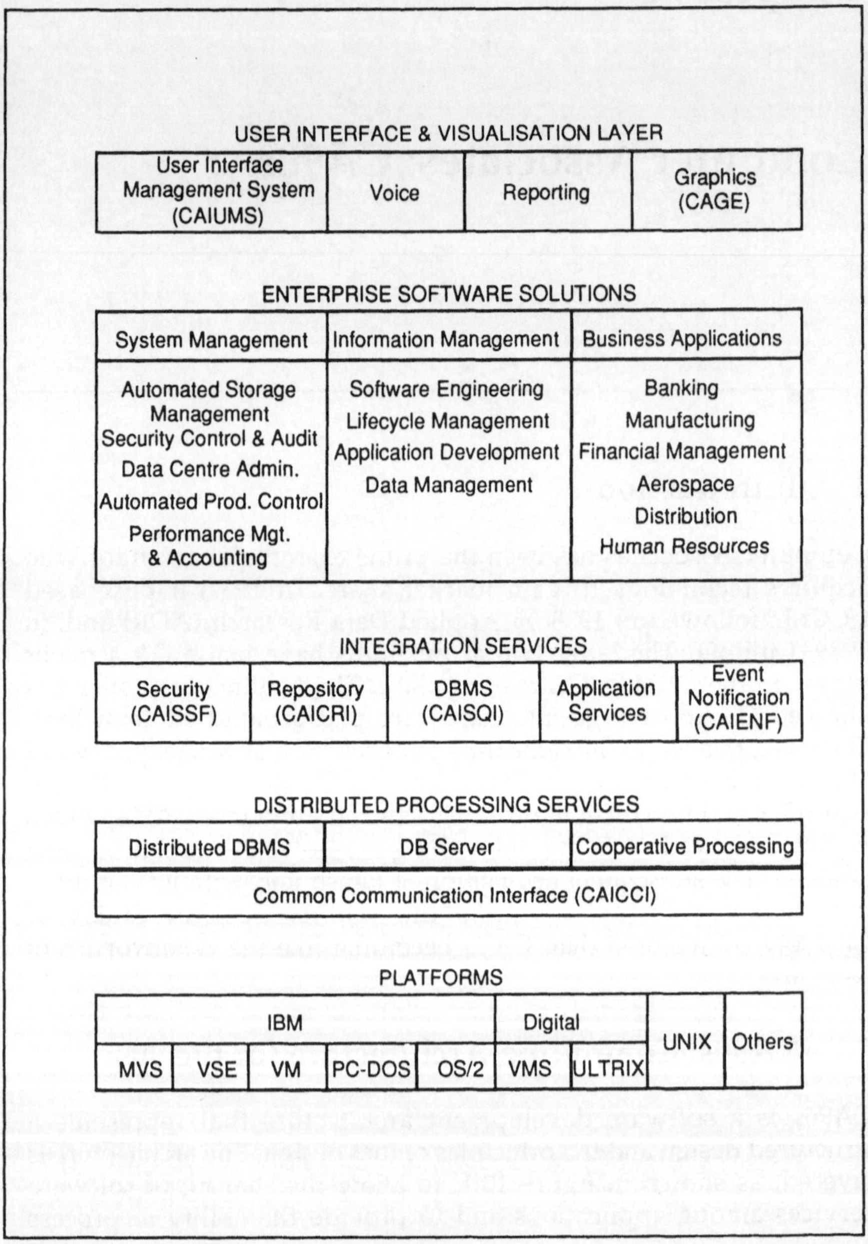

Figure 13.1 CA90s Architectural Layers

information across an organisation. CA90s is based on the technology that resides in over 200 of Computer Associates' existing products.

2.1 Guiding principles of CA90s

The guiding principles for CA90s are as follows:

- to preserve clients' investment in hardware, software and expertise, while enabling them to take advantage of emerging technologies
- to endorse and enhance established industry architectures and standards (such as IBM's SAA and Digital's NAS)
- to promote the integration of software solutions across product families, operating system and hardware platforms
- to enhance and extend distributed processing across multiple operating system and hardware platforms
- to facilitate the portability of software solutions, engineered for efficiency on each platform.

2.2 CA90s basic architectural layers

The basic layers of the CA90s architecture are:

- enterprise software, encompassing system management, information management and business applications
- three categories of services:
  - user interface and consistent 'visualisation'
  - integration
  - distributed processing
- multiple platform support, allowing solutions to extend across a range of operating systems and hardware platforms.

2.3 Enterprise software solutions layer

The enterprise software solutions layer consists of applications that address the automation of information processing. These are defined in three functional areas:

- system management software – integrated applications that automate the data centre operation

- information management software – including database management, software engineering, application development and life cycle management
- business application software – end user computing solutions.

2.3.1 System management software solutions

The system management software solutions include:

- automated production control, providing the means to manage the complete production workflow
- storage and resource management, providing multi-media, multi-operating system storage management of data
- security, control and audit, addressing the requirements of asset protection including access control, integrity checking and violation monitoring
- performance management and accounting, providing service level-oriented performance management and tuning, charge-back and accounting for information systems resources and multi-platform capacity planning
- data centre management, providing problem, change and financial management across user networks, with integrated configuration and inventory control.

2.3.2 Information management solutions

The information management solutions aim to improve productivity by integrating software technologies. These solutions include:

- database management, providing repository, DBMS, support for a common access method, and data and process distribution
- software engineering, including third and fourth generation languages, non-procedural techniques, application generators, knowledge-based systems and information delivery
- lifecycle management, including business modelling, analysis and design, data modelling, project management, configuration management and an environment for developers.

2.3.3 Business application software solutions

The business application software solutions address both vertical and horizontal markets, and include products for:

- financial management
- human resource management
- banking
- manufacturing management
- distribution management
- decision support
- visual information (CAD/CAM & DTP).

2.4 Service layers

2.4.1 User interface and visualisation services layer

The user interface and visualisation services layer is intended to help the end user. By providing a consistent view across platform and job functions, it acts as the interface between the human and the machine. This layer includes:

- user interface management
- graphics
- reporting
- voice.

These services are supported on, and allow compliance with, IBM's CUA, Digital's DECWindows, Apple Macintosh and UNIX platforms.

2.4.2 Integration services layer

The integration services layer insulates the enterprise software solution from operating system dependencies. This layer also supports overall integration among solutions. Among the services provided are well-established CA techniques such as database management, security, project management, change control and expert systems. Two new technologies are also included: repository and event notifications. The CA repository will serve as the vehicle for integration of a company's information needs. The CA event notification facility

(ENF) provides a single shared interface that is consistent across operating systems and hardware platforms.

2.4.3 Distributed processing services layer

The distributed processing services layer provides a number of capabilities, sharing a common communication interface; these include co-operative processing, database server facilities and distributed database management.

2.5 Multiple platform support

From a position of strength in the IBM world – its product range provides support for all IBM platforms from MVS to PS/2 – CA moved, via the acquisition of Cullinet, firmly into the Digital camp.

CA's stated direction in the 1990s is towards UNIX, and we have already seen a few products for this platform. Additionally, CA is a UNiSYS strategic partner, with SuperProject as a preferred project management tool in UNiSYS environments.

More fundamentally, though, CA has demonstrated a commitment, wherever appropriate, to basic standards. This should ensure that it has the ability to implement products on any platform.

3 CA90s information management services

The information management services software (see figure 13.2) addresses the principles and architecture of CA90s by providing data management and platform independent software engineering technologies. The software also forms the basis for the data management services of the CA90s integration services layer.

This dual capacity provides CA customers with the means to develop and enhance their own applications. The information management software covers the data management functions of DBMS and repository support, as well as providing software engineering capabilities, including lifecycle management and systems for application development.

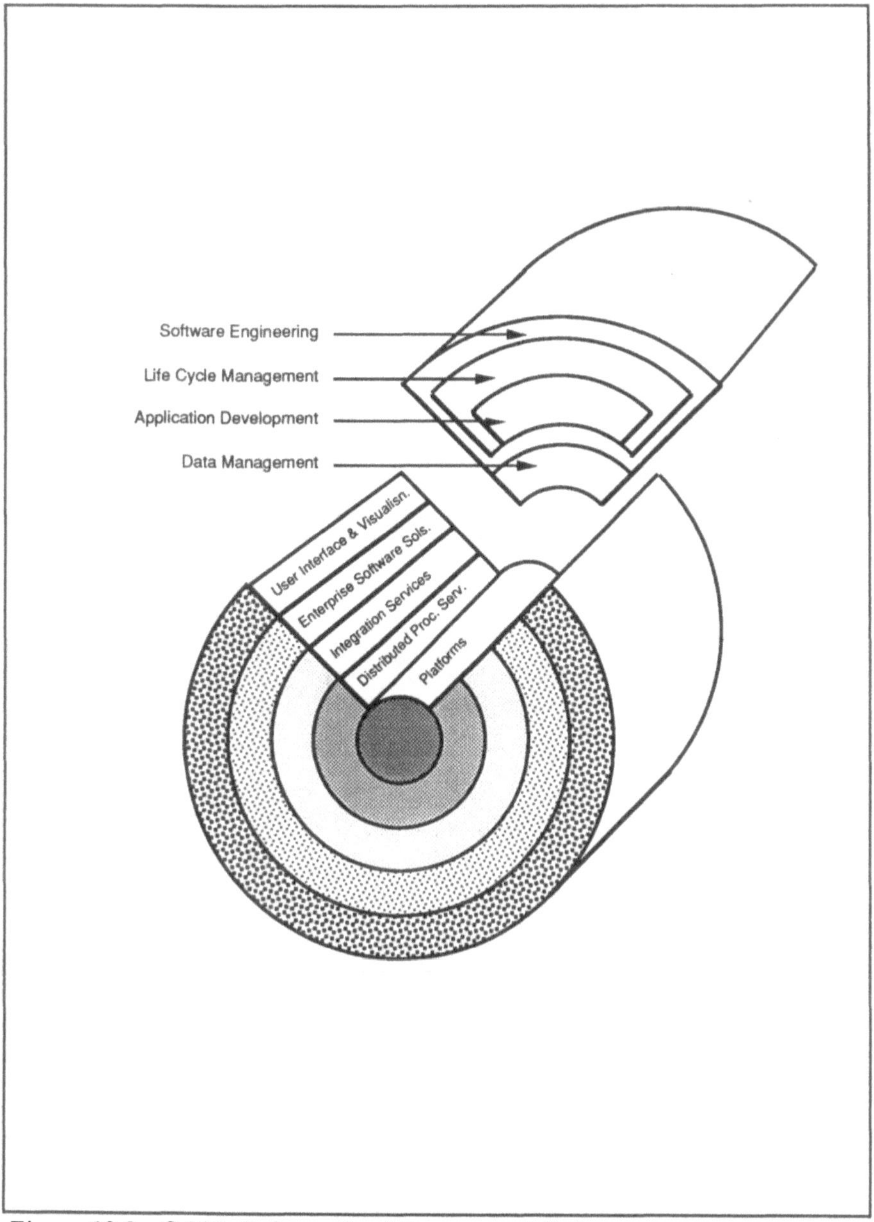

Figure 13.2 CA90s Information Management Software

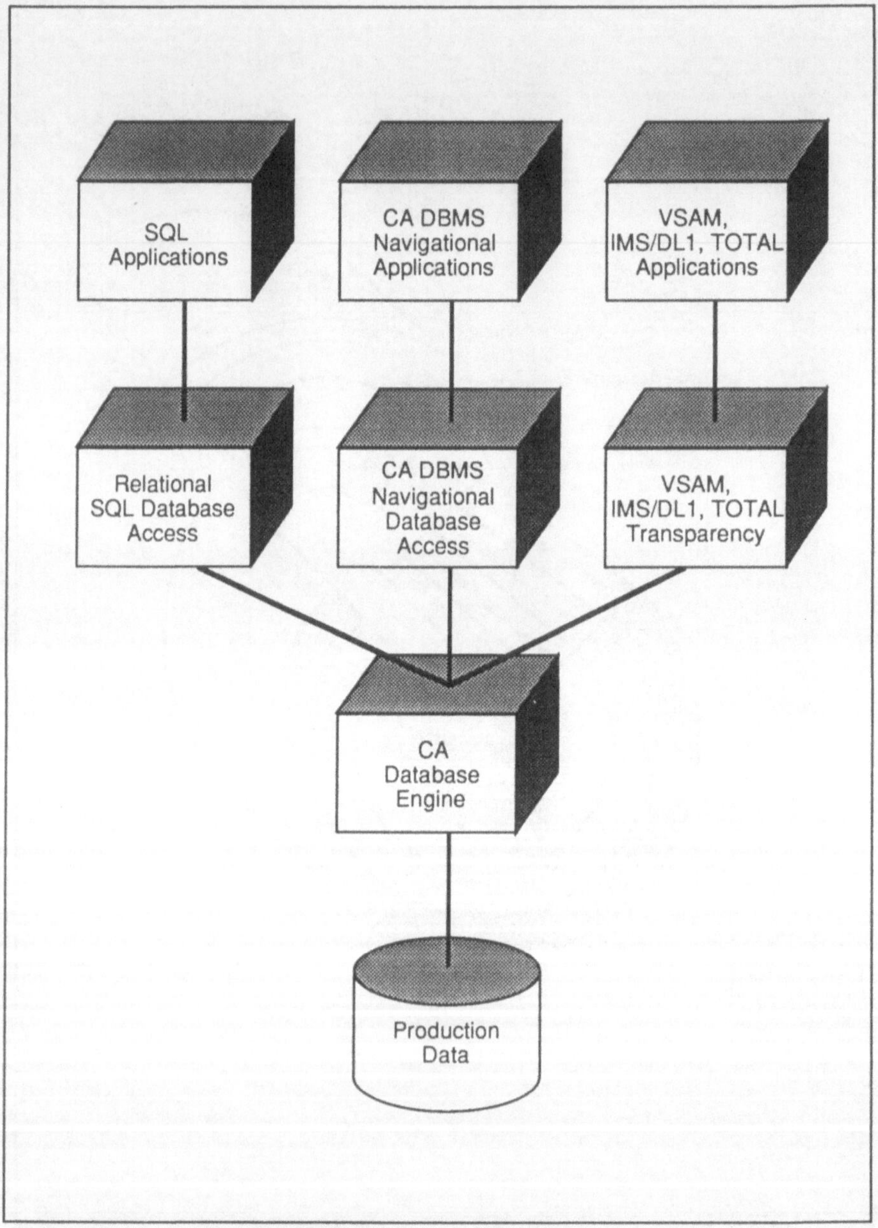

Figure 13.3 CA Database Management Systems

3.1 Database management systems

CA provide for multiple data access techniques within a single database. This single database strategy is achieved through the addition of SQL technology to proven CA DBMS products, as well as through the use of transparency software. The latter is supplied for VSAM, IMS and DL/1, DB2 and TOTAL. The database technology supports simultaneous access to data through the transparency service, and SQL or other database access methods. The database management services (see figure 13.3) provide the following functionality:

- data management, providing multiple views of the same information through standard query and reporting services
- data sharing, allowing applications to reference and rely on a common set of information
- data integrity and security, ensuring the accuracy and reliability of critical information
- data availability, enhancing the availability of information to end users through dynamic database maintenance capabilities, and through distributed database services
- performance tuning, regardless of access technology
- portability across multiple operating systems and hardware platforms
- distributed information, allowing organisation-wide applications to co-operate across a mix of platforms, with full data integrity.

CA has a great degree of experience in SQL technology. Its extended SQL includes the current SQL 1.1 standard as a subset. It also supports both the Federal information processing standard (FIPS) variant of the ISO standard, and IBM's SAA SQL standard.

The database management services support an open architecture. The CA SQL interface (CAISQL) provides a standardised interface to several SQL-based RDBMS products, and supports both dynamic and static SQL. This API is shown in figure 13.4.

Alternative database access methods are under consideration for support; these include:

- entity-relationship models, in particular the model used in the ANSI IRDS standard as a means of definition and manipulation of repository data

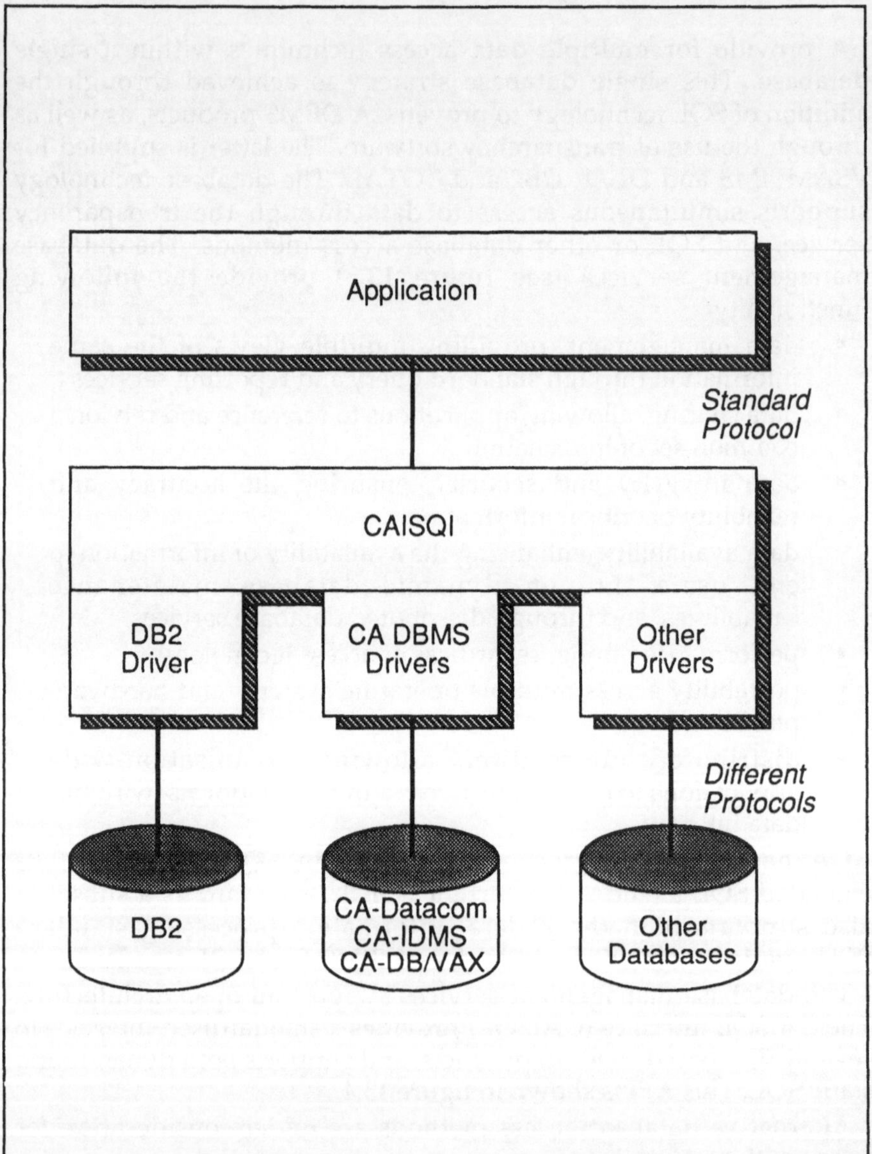

Figure 13.4 The CAISQL API

- multimedia, to support complex, unstructured objects such as text, documents, graphics and images
- semantic and object-oriented methods.

3.2 Repository support

The CA repository services (see figure 13.5) are based on the technology in their data dictionary products, CA/Datadictionary and CA/IDD. These services allow users to manage and control all the information of an organisation, for information management purposes as well as for business information using CA's systems management and business applications software. It is compatible with IBM's RM/MVS repository manager and Digital's CDD/Repository. The repository services play an active role in the CA90s integration services layer, supporting all CA enterprise software solutions.

The repository services encompass three main levels:

- an API known as the CA common repository interface (CAICRI), which handles the interaction between users and the information model
- the information model, which describes the information structure as well as the external views utilised by the API and storage views
- storage management facilities, to handle the actual storage of the data.

The CAICRI is a full-function, program-call interface to CA repository services, allowing CA software and user programs to populate, navigate and maintain the information under the control of the repository. The interface promotes a common view of an organisation's information, even where, physically, this may be described in multiple dictionaries. CAICRI supports IBM's SAA common programming interface (CPI) extension for repository services, thus allowing any third party vendor product to interface to CA repository services. CA has stated that CAICRI will also support the ANSI IRDS services standards, should they differ from IBM's implementation.

All information under the control of the repository is described in the repository services information model (see figure 13.6). This model is used as an information interface, providing a view of the information structure controlled by the repository. The CA

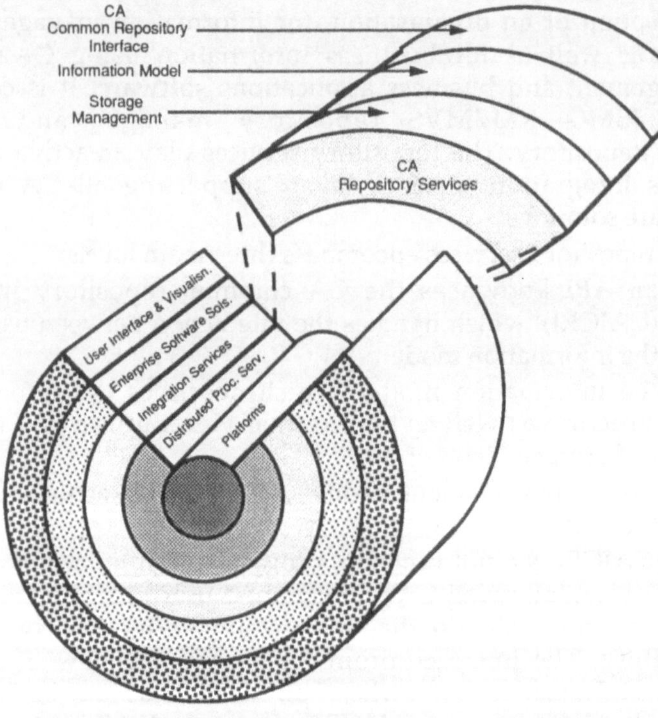

Figure 13.5 The CA Repository Services

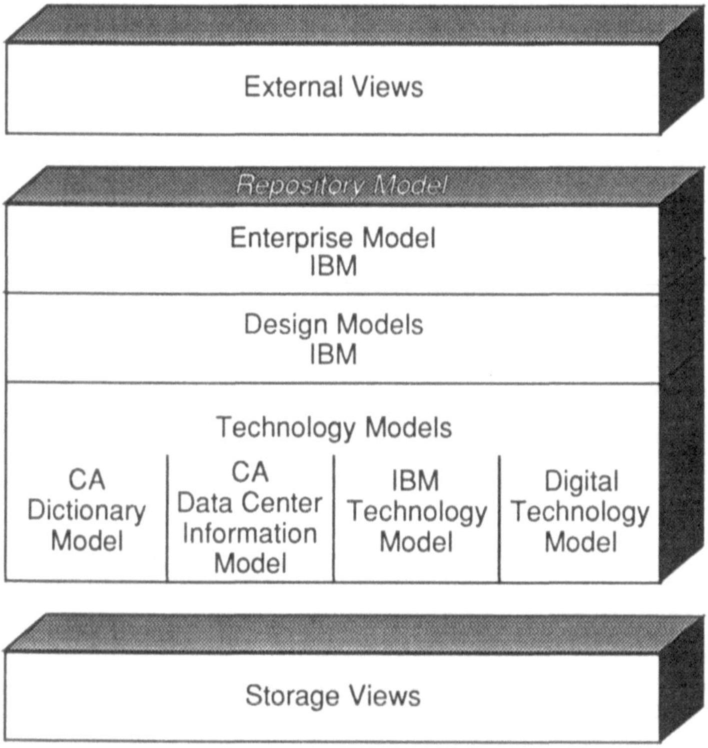

Figure 13.6 The CA90s Repository Services Information Model

information model follows the ANSI/SPARC three-level model, and consists of:

- the external views, used by the API as the means of communication and defining the unit of data transfer
- the repository model, defining the various entity types supported by the repository
- the storage views that describe the underlying storage management.

The repository model is the conceptual schema that defines the data 'in-and-of-itself' – an implementation-independent view – regardless of usage or storage. It is based on the entity-relationship modelling techniques and applies to three models:

- the enterprise model, which defines the structure of the organisation, its business rules, its policies, its procedures and its data
- the design model, which describes the design of the automated solutions of the organisation
- the technology model, which defines the implementation of the solutions in terms of the database and application definitions.

The data centre information model defines the following types of information for CA's entire enterprise software solutions line:

- rules
- product processing options
- product information.

In terms of rules, many CA products, such as scheduling and security software, are rule or policy based. The user defines the set of rules that determines how the software will handle different processes and actions. These rules are currently recorded in product files or databases.

Product processing options are provided by most CA products. These indicate how the software will operate under specified conditions. These options are currently supplied by means of parameters or control file entries.

In order to supply product information, most CA products use additional file and databases to record information specific to that product, as part of their processing. Usually this information takes the form of logging, audit trails, and recovery information. This category also includes the information necessary for event sharing through the CA event notification facility.

The repository services offer a variety of storage management techniques, depending on the needs of each functional area. The four categories of storage management supported include:

- dictionaries, including CA products, IBM's RM/MVS repository manager and Digital's CDD/Repository
- DBMS support, including CA products, IBM's relational DBMS products (DB2, SQL/DS, SQL/400, and DBM), and Digital's Rdb
- catalogues, often used in system management products, which support simple data structures, and include standard operating system files such as VSAM and BDAM, interfacing being handled by native access
- memory-resident information, often used in system management products to provide very fast access to real time information; access is handled via the services delivered by each product.

Query and reporting facilities are necessary for any repository. At the level of any given dictionary, native dictionary services, query facilities and reporting services are available. In addition the CAICRI, a forms-based query and reporting facility, is provided as part of the repository services. Through non-procedural, fill-in-the-blank methods, users are able to define forms that can be used as panels for online queries or as layouts for reports.

3.3 Life cycle management

Effective life cycle management, in CA's view, helps to automate and integrate the following tasks:

- project management
- front end CASE solutions
- configuration management.

3.3.1 Project management tools

CA currently provides a wide range of project management tools, based on a variety of platforms ranging from PCs to mainframes,

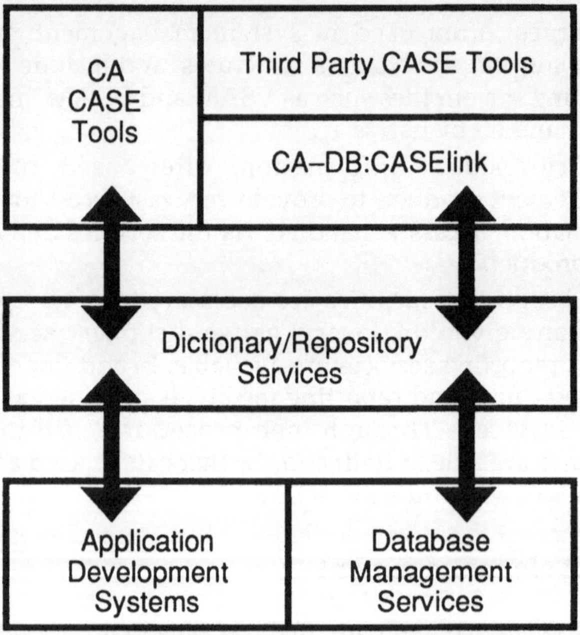

Figure 13.7 CA Front End CASE Integration

with the capacity to transfer information between them. These tools provide users with automation facilities for estimation, planning and tracking of projects, and management facilities for controlling tasks, schedules, staff, resources and costs. Under CA90s, these tools will utilise the graphics and reporting services already defined in the architecture. Integration with other CA90s services will enable direct connectivity of project management tools to CA's database management and repository services. Current CA products are ESTIMACS, PLANMACS, TELLAPLAN and SuperProject.

3.3.2 Front end CASE

Front end CASE has always been the weak spot in CA's product portfolio. With the acquisition of Cullinet, they obtained IDMS Architect, a version of LBMS' Automate Plus. The acquisition of ADR brought them Depictor, a product only just released at the time of the purchase. CA really had no visibility in this product area before CA90s. Today, the company refers to its front end CASE solutions without specifically naming any products.

CA talks about the integration of its workstation-based graphical design tools for database and process design with dictionary models and back end CASE tools (see figure 13.7). However, what is important is the description of how CA's open architecture will support the full integration of third party products with the database management services, repository services and application development systems (see below, section 3.4).

February 1992 saw CA release CA-Conceptor, a tool for designing and building applications for Datacom/DB, IDMS, DB2 and CA-DB UNIX. It produces database definitions from user diagrams of applications. However, CA-Conceptor, a recent re-badging of a French product – IBSI's Conceptor – is unlikely to find a ready market on a worldwide basis, as it can support only the Merise and Yourdon approaches to analysis and design.

3.3.3 Configuration management tools

CA supplies many products in the configuration management arena and, through acquisitions, has been a pioneer in library management and change control. CA90s demonstrates a realisation that applications comprise many components; these include program modules, databases statements, SQL plans, schemas, and subschemas, reports, panels, help text and messages. All these components must be managed during the life cycle of the application as it is planned, analysed, designed, built, tested, corrected, released and maintained.

In the integration services, CA90s includes a change management service (see figure 13.8) that manages and controls change requests, in conjunction with the repository services, through which all objects are defined, and CA's products, which manage the objects. This service allows all changes to the information processing operation to be recorded, approved and implemented in a consistent way. In addition, it provides information about the impact of any change on other objects. Current CA products include LIBRARIAN.

3.4 Application development systems

CA's information management services provide a comprehensive set of integrated, portable and open application development systems (see figure 13.9). Capabilities are provided for on-line, interactive development environments, using which developers can move through the development process. These facilities include testing, debugging, generation of test data, and optimising code. The application development systems are integrated with the repository services, library management systems and the CA standard security facility. They include the following types of tools to meet the varied organisational development needs:

- fourth generation languages – ADDS, cBASE and IDEAL
- code generators – TELON
- non-procedural, rule-based systems – DB:Expert
- expert systems – DB:Expert
- query systems – DataQuery
- COBOL development environments – DB:Generator, MetaCOBOL, Optimizer and PROMACS.

These tools will work not only with CA products but also with DB2 in an IBM environment and with Rdb in a Digital VAX/VMS environment.

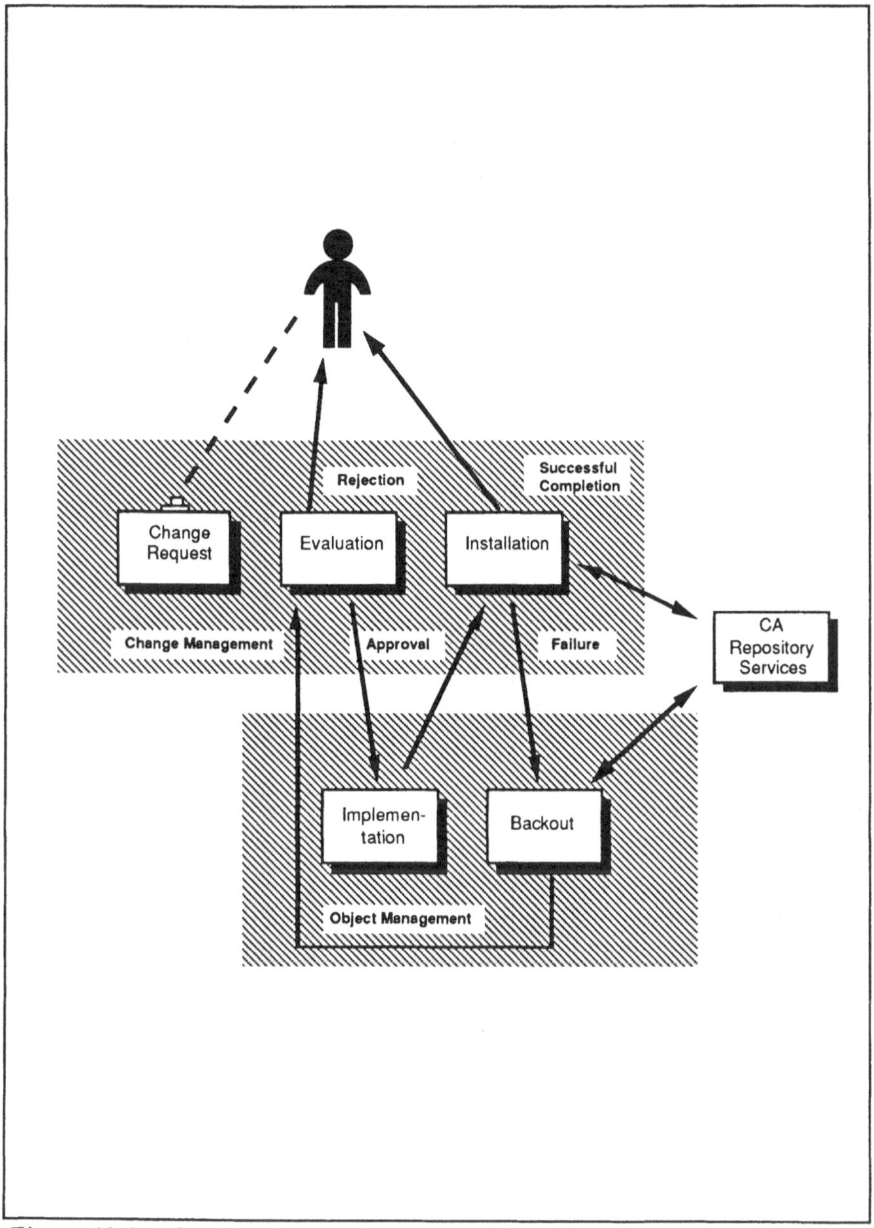

Figure 13.8 The CA Change Management Service

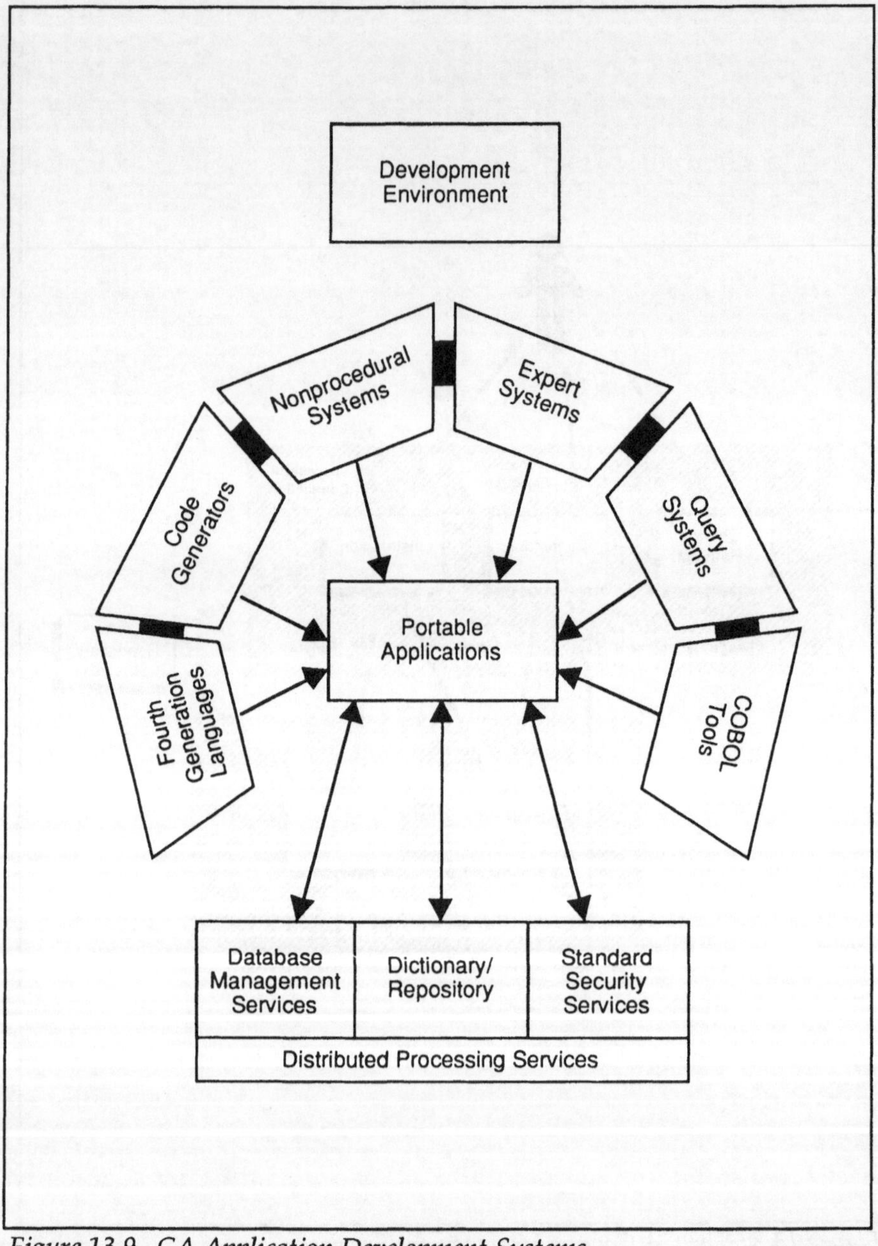

Figure 13.9 CA Application Development Systems

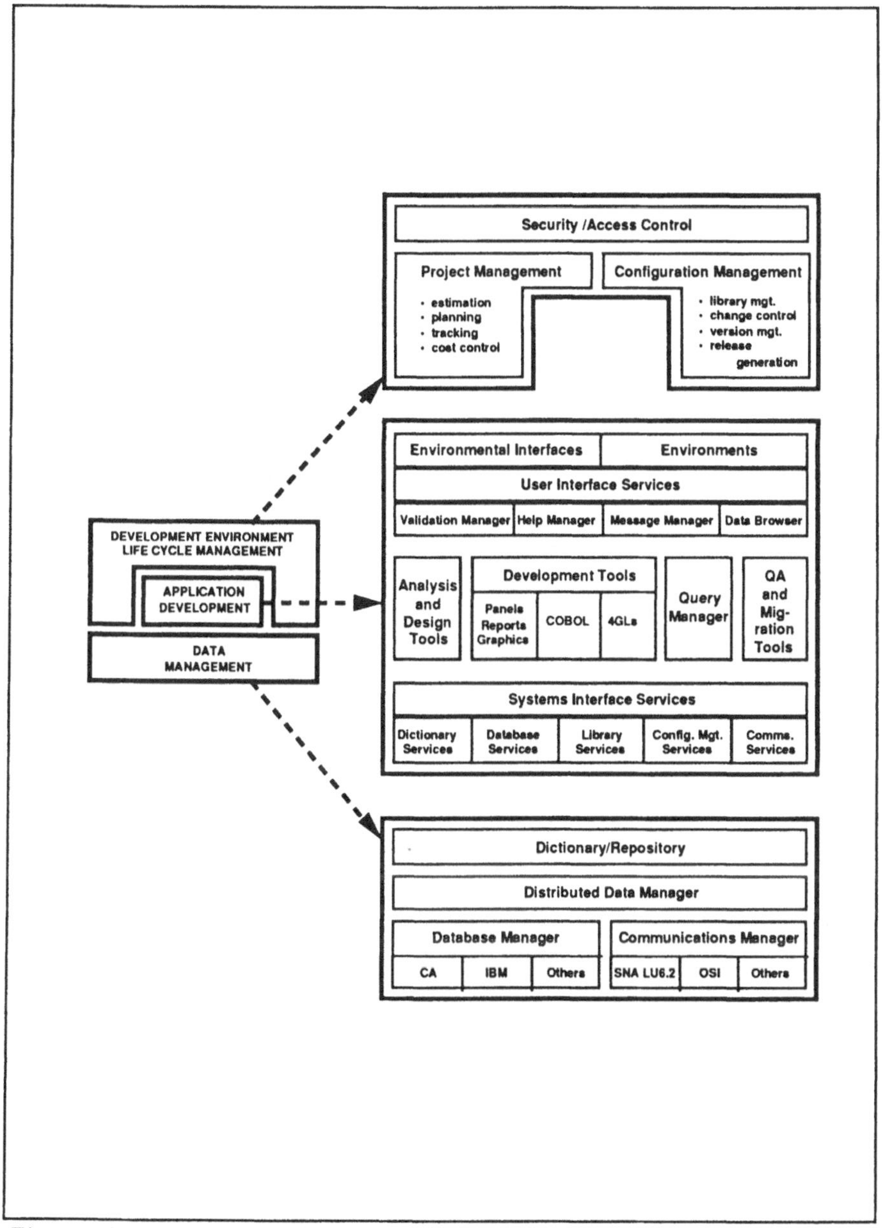

Figure 13.10 CA-ACE Architecture

4 CA-ACE: Application Construction Environment

CA-ACE was the fore-runner of CA90s, but it is useful to review and understand it, as it was the foundation for the enterprise software solutions. CA-ACE had, as one of its aims, the integration of relational and navigational database access with common operation, integrity and security management, on multiple platforms – in other words, the integration of ADR, Cullinet and CA products.

The architecture for CA-ACE consisted of an integrated system of PC and mainframe tools for constructing applications based on relational database management systems. These tools were aimed at improving productivity across the entire application life cycle, providing facilities to deal with the requirements of different types of applications, while meeting the performance requirements of production systems.

The overall information system architecture (shown in figure 13.10) had three major components:

- life cycle management, an integrated development environment providing complete development life cycle support
- application development, a system of application development tools supporting COBOL, high level languages and ad hoc queries
- data management, a relational DBMS with support for SQL, a repository and distributed databases.

There was another dimension to this three-level architecture; the computing platform, covering computers and operating systems. CA-ACE's platform independence and distributed data management laid a foundation for application portability, development workstations, and distributed processing. This flexibility was intended to preserve investments in application development, since it allowed computer hardware and system architectures to be changed as business needs and technology evolved.

5 Conclusion

CA has still to announce a convincing front end CASE product. Depictor, the ADR product that was in beta release at the time of acquisition, was put on the shelf. The Cullinet version of LBMS' Automate Plus suffers from a double problem: it is based on an old version of a tool that is now obsolete in any event – see Chapter 18. Both ACE and CA90s talk about CASE tools, and CA is particularly strong in the CASE infrastructure tools area, with configuration management, project management and verification, validation and testing products. The other strength for CA is with back end CASE, with tools such as IDEAL (ex-ADR), ADDS (ex-Cullinet) and cBASE. However, the purchase of Pansophic, early in 1992, has complicated the picture further.

CA has the technology to produce a viable development environment for any user. Customer uncertainty is all too likely, given the amount of choice, with numerous competing products in the same field. For some, also, it must be said that the company persona puts people off. We feel that CA has developed a fine architecture but that it is now time to deliver. Its continual round of purchasing is confusing the picture, while failing to deal with the current weak spot, namely front end CASE. CA-Conceptor, with its specific Merise and Yourdon support, does not seem like the answer, and CA still appears somewhat exposed.

6 References

1 *Computing* (1991), *CA Aims at Pansophic Buyout for CASE Entry*, 12th September 1991

2 CA (1990), *CA90s Executive Overview*, SB05ZZ CA90E

3 CA (1990), *CA90s: Computing Architecture for the 90s*

4 CA (1989), *CA-ACE, Statement of Direction*, February 1989

5 CA (1989), *Statement of Direction*, from *Database Management and Application Development on Multiple Platforms*, September 1989

6 CA (1990), *CA-IDMS/DB and CA-IDMS/DC* (update to previous publication), February 1990

7 CA (1989), *Application Development Systems*, CA Position Paper, February 1989

8 CA (1989), *COBOL Development Environment*, CA Position Paper, February 1989

9 CA (1989), *Multiple Platforms Support*, CA Position Paper, February 1989

Chapter 14

Software AG: Integrated Software Architecture and ENTIRE

1 Introduction

Software AG has for many years sold its products across 3 different hardware platforms: IBM, Digital and Siemens. Best known for its DBMS product – ADABAS – and its 4GL product – NATURAL- it was a late entrant into the dictionary market, not providing a product – PREDICT – until 1983. Software AG never entered the relational database market, stressing instead the object oriented virtues of the well-established ADABAS and NATURAL. This, it believes, puts Software AG in the driving seat for the 1990s.

2 Integrated software architecture (ISA)

In answer to IBM's SAA announcement, Software AG produced ISA (integrated software architecture). This took the concepts of SAA and applied them to the company's products across the three hardware platforms, integrating fourth generation technology and basic functions in a visible manner. The company believes that this will yield four major advantages, by:

- allowing users to customise the environment
- allowing developers and users to extend the environment
- providing system independent interfaces to many functions
- using accepted industry standards for its integration of architecture, operating environments and protocols.

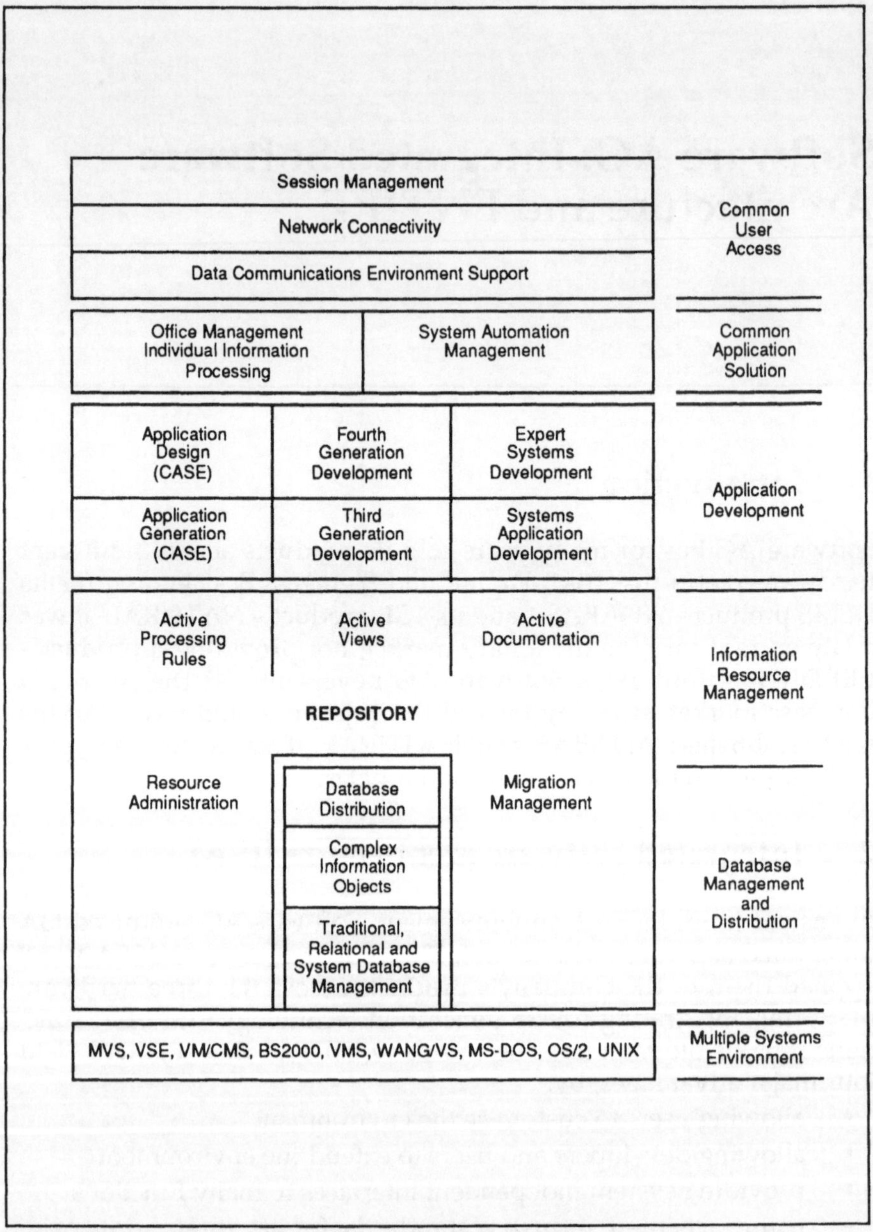

Figure 14.1 Software AG's ISA Architectural layers

The basic principle behind ISA is the separation of application logic (the modelling of operational knowledge and business processes) from physical access (management of a wide variety of hardware and system software configurations). This, in Software AG's view, allows organisations to concentrate on fulfilling their strategic, tactical and operational business assignments. ISA achieves the independence of system environments and the portability of IS/IT applications through an architecture based on integration and openness.

Integration is achieved through two principles:

- one piece of information in one place
- individual functions which can be combined at will.

Openness is used to enhance integration in two ways:

- the interfaces, conventions and protocols linking the various function layers are documented and guaranteed
- the interfaces, conventions and protocols that develop with industry standards are integrated into ISA's productivity.

ISA achieves its openness and integration by separating functions structurally into a number of well-defined software layers:

- common user access
- common application solutions
- application development
- information resource management
- multi-systems environment support.

Figure 14.1 shows these five ISA layers.

The ISA architecture describes internal aspects such as register sets, processor status and command sets as well as external features such as networks and multi-level architectures (mainframes, departmental computers and workstations). Hardware architectures form the basis for the implementation of products such as operating systems and network software. Software architectures are collections of interfaces, conventions and protocols reflecting the allocation of functions to different levels of abstraction and independence.

2.1 Multiple systems environment layer

ISA products are supported on a range of operating systems and hardware platforms. These include mainframes, departmental computers and workstations. Hardware platforms and operating systems supported include:

- IBM:
  - MVS
  - DOS/VSE
  - VM/CMS
  - MS/DOS
  - OS/2
- Digital:
  - VMS
- Siemens:
  - BS2000
- Wang:
  - VS.

In August 1991, ADABAS and NATURAL were released for UNIX, running under Hewlett Packard's HP-UX on the 9000 Series 800 and NSC's SINIX on the MX300 and MX500 machines.

2.2 Information resource management layer

ISA products support the storage, management, manipulation, selection and analysis of data, text and other information objects. These functions can be performed using ADABAS, VSAM, IMS/DB, DL/1, DB2, SQL/DS, SESAM, RMS, Rdb or operating system data. ISA uses a semantic database to implement the conceptual model; but at the same time supports the SQL standard as a means of access and definitions. Active, integrated control of data is provided by Software AG's repository product, PREDICT.

The information resource management layer is split into sub-layers:

- information resource management technology
- database management and distribution technology.

The layer contains the following products:

- elements of PREDICT:
  - the integrated active dictionary
  - PREDICT Application Control (PAC), a controlled environment to ensure the integrity of an application throughout its life cycle

- elements of ADABAS
  - the database management system itself
  - DL/1 Bridge
  - Fastpath, to optimise computer resources for ADABAS
  - Text Retrieval, to manage unformatted text using ADABAS
  - TPF (transaction processing facility), to control database and NATURAL dialogues for high volume transaction processing
  - HPE (high performance environment), an MVS/XA, MVS/ESA or VM facility that serves as an operational shell for running Software AG products in a single address space
- NATURAL Process, providing system data management for system programmers
- Net-Work, providing a distributed processing capability in a heterogeneous environment
- Review, a performance monitoring and accounting system.

2.3 Application development layer

ISA products offer an integrated system of methods and tools for the development of applications. The heart of this layer is NATURAL, the fourth generation environment product. Software AG has developed a number of add-on products to NATURAL, including:

- NATURAL Architect Workstation, an analysis and design workbench for the Macintosh
- NATURAL Construct, an application generator
- NATURAL for DB2, SQL/DS, VSAM, DL/1, OS/2, Rdb, RMS, SESAM and UNIX
- NATURAL Expert, an expert system development environment
- NATURAL Graphics, a business graphics interface for NATURAL
- NATURAL ISPF
- NATURAL Optimizer Compiler, a performance enhanced compiler for NATURAL
- NATURAL Security, the security system

- PREDICT CASE, a multi-user system for the planning, documentation and development of application systems
- PREDICT CASE Workstation and Coordinator, a PC-based graphical user interface to an I-CASE application development environment.

The application development layer contains two Software AG front end CASE products. PREDICT-CASE is an IBM PC/mainframe product, while NATURAL Architect is a Macintosh-based tool, designed for the DEC environment.

2.4 Common application solutions layer

ISA products offer a range of standard solutions for applications, information processing and office function. Software AG's products in this layer are based on NATURAL and include:

- Super NATURAL, a menu or SQL-driven query and report generator for business users
- NATURAL Graphics, a graphics output editor
- NATURAL Spreadsheet, a three-dimensional spreadsheet
- NATURAL Statistical Link, providing integration with SAS products
- NATURAL W Link, providing integration with Comshare products
- NATURAL Document Manager, a free text retrieval product
- NATURAL Operations, a production and scheduling control system for operations
- NATURAL Console, an on-line facility for controlling operating systems
- NATURAL Output Management, providing output management for the data centre
- NATURAL Elite, a computer-aided training package
- CON-NECT, an office automation system.

2.5 Common user access layer

All ISA users communicate with the system using a uniform interface. Their communication can be based either on workstations or on a number of different TP monitors, such as COM-PLETE, CICS, TSO IMS/DC, CMS UTM or VMS. Products in this layer include:

- NET-PASS, a session manager for VTAM networks
- VM-PASS, a session manager for VM networks
- NATURAL Multipass, a session manager for NATURAL under BS2000
- NATURAL Connection, a window-based product which provides micro-mainframe integration for Macintosh and PC LANS
- COM-PLETE, a conversational TP monitor.

2.6 ISA summary

For all its claims of openness, ISA is a closed environment. There are few links to third party products – only Comshare's System W, SAS and LBMS' Automate Plus (now an obsolete tool) are documented. ISA is tightly coupled around Software AG's own product set, with NATURAL playing a key role as the fundamental common product.

3 ENTIRE: Software AG's function server technology strategy

With ENTIRE, Software AG has introduced, within an open ISA framework, a client-server technology solution that spans multiple platforms including existing mainframe environments. ENTIRE technology provides an architecture that bridges the transition from a solely mainframe-based environment to a mixed client-server environment based on mainframes, UNIX systems and PCs.

The goals of the ENTIRE function server strategy are as follows:

- to make very different functions look as similar as possible
- to make services accessible via standard SQL commands wherever possible
- to present server functions in an object-oriented fashion for easier understanding, reusability and consistent message-based communication
- to establish a consistent logical protocol for communicating requests between clients, agents and servers.

The ENTIRE function server technology consists of six essential ingredients:

- open communications:
  - function server protocol
  - open communication network
- clients:
  - portable application agents from Software AG, users and third parties
- servers:
  - cooperative application server
  - open SQL server
  - open function server.

3.1 ENTIRE function server protocol

The critical success factor in a client-server environment is communications. The keystone here is a protocol which can satisfy both today's and tomorrow's information distribution demands. The ENTIRE function server protocol (FSP) describes the rules and conventions to be obeyed when requesting function services. High-level requests are made available for:

- database requests through SQL, providing open SQL access across heterogeneous SQL DBMSs
- extended database requests for text, images, structures, geographic and voice through an extension to SQL
- information object requests
- electronic mail requests through SNADS or X.400
- electronic data interchange (EDI) requests
- remote procedure calls through RPC
- user-defined, high level function requests
- document requests through DCA or ODA.

3.2 ENTIRE open communication network

The ENTIRE open communication network (see figure 14.2) is the interoperability backbone of Software AG's client-server technology. It provides a logical communications layer above existing network

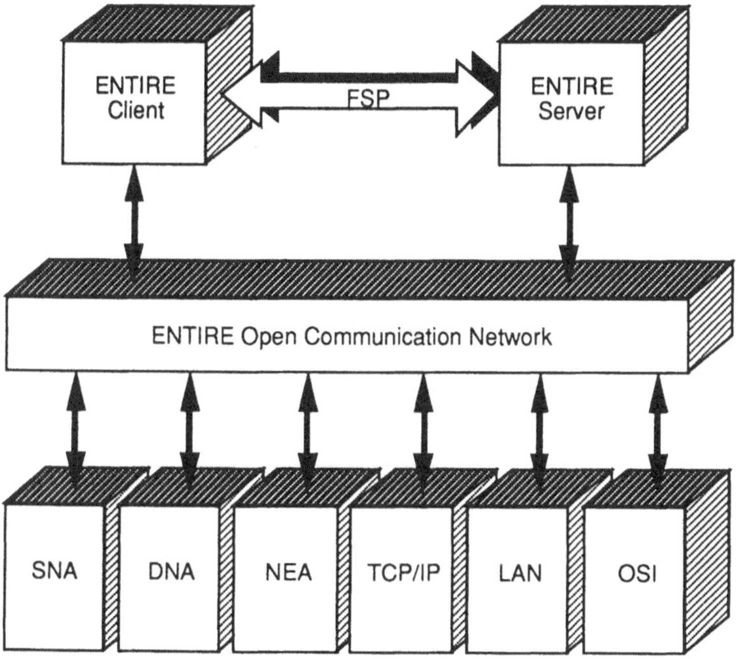

FSP – Function Server Protocol (Software AG)

SNA – Systems Network Architecture (IBM)

DNA – Digital Network Architecture (DECnet)

NEA – Network Architecture (Siemens–Nixdorf Transdata

TCP/IP – Transport Control Protocol / Internet Protocol (DARPA)

LAN – (including NOVELL, LAN Manager, LAN Server, BANYAN)

OSI – Open Systems Interconnect (ISO)

Figure 14.2 The ENTIRE Open Communication Network Architecture

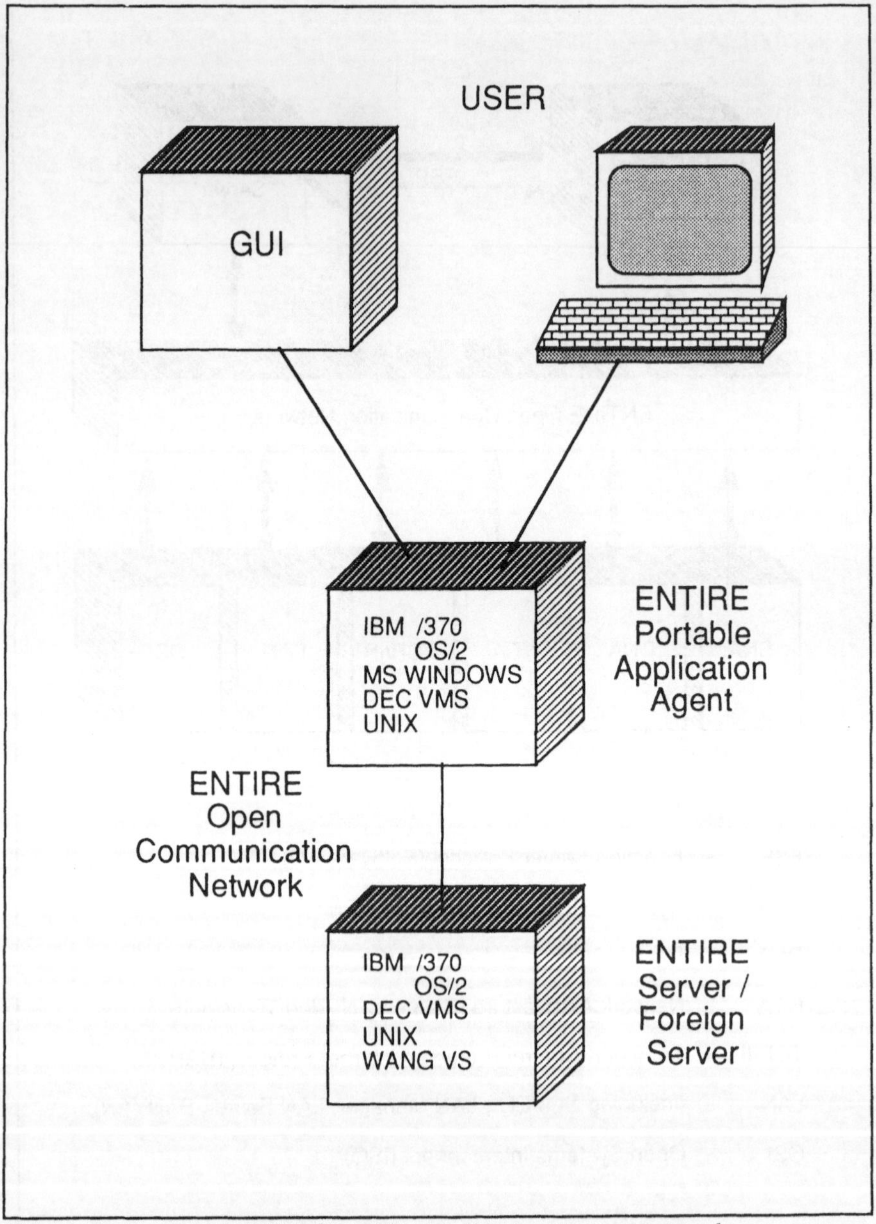

Figure 14.3 The ENTIRE Portable Application Agent Network

architectures, so that WANs and LANs can be integrated. This component can work with IBM's SNA (and in particular LU6.2, the key SAA communications protocol), Digital's DECnet, TCP/IP for UNIX and, in the open systems world, the OSI standards for RPC, TP and RDA. The ENTIRE open communication protocol acts as the transportation medium for the ENTIRE function server protocol.

3.3 ENTIRE portable application agent

The ENTIRE portable application agent, based on NATURAL, provides a fourth generation application development and execution time environment (see figure 14.3). Its implementation objectives are focused on productivity, performance and integration.

In the role of application client, it supports both GUI and 'dumb terminal' interfaces. Support is provided for IBM's SAA/CUA standards, including both Presentation Manager and Microsoft's Windows, as well as for both the X-Windows implementations, OSF/MOTIF and OPEN LOOK. In addition, support for the Macintosh GUI is provided.

In the role of application server, it provides access to commercial DBMS products such as IMS, DL/1, DB2, Rdb, ADABAS and other SQL-based products. It also supports on-line transaction processing with common TP monitors such as CICS, IMS/DC and ACMS.

3.4 ENTIRE cooperative application server

In a client-server environment it may well be inefficient to simply ship database requests. It is often better to locate application functions with the information resources, in order to minimise network traffic. The ENTIRE cooperative application server (see figure 14.4) provides the means to distribute functionality across all the clients and servers which it supports. An application can be cut into pieces, each of which can be processed in any appropriate hardware environment connected by the ENTIRE open communication network.

3.5 ENTIRE open SQL server

The ENTIRE open SQL server (see figure 14.4) provides access to heterogeneous SQL databases from different vendors. It provides the

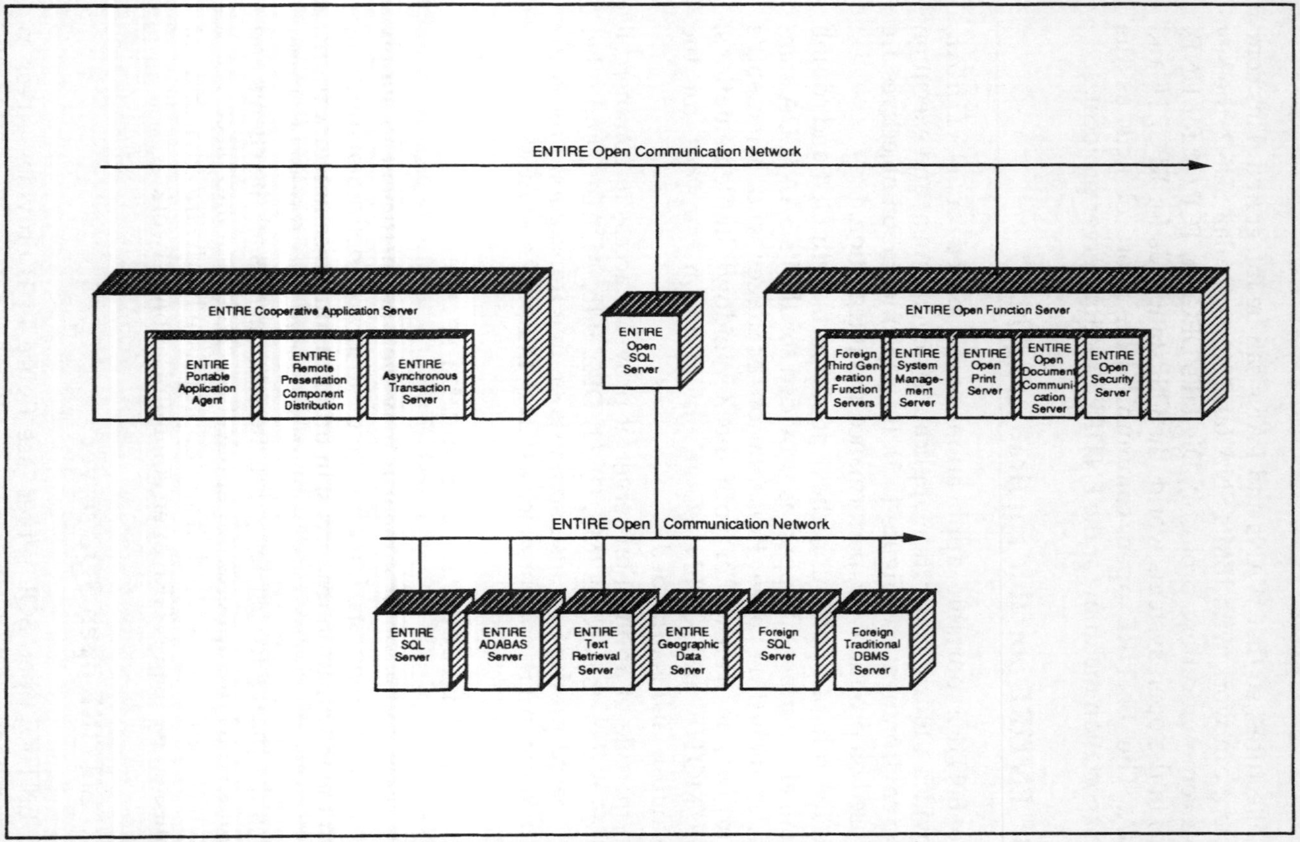

Figure 14.4 ENTIRE Servers and the ENTIRE Open Communication Network

engine which drives the database server, and supports the ANSI SQL 1.1 standard.

3.6 ENTIRE open function server

The ENTIRE open function server (see figure 14.4) provides the means to integrate third generation language functionality, as a server, into the ENTIRE function server world. It provides an execution time environment for traditional applications written in COBOL, PL/1, Fortran or C, and can be used to interface the ENTIRE open communication network to existing implemented applications.

4 Further reading

1 Software AG (undated), *ISA: The Software Architecture – Strategy for Success*, ISA 1088427GB
2 Software AG (1989), *Product Profits*, SAG-PRF-001, Nov 1989
3 Software AG (undated), *Product Catalog*
4 Pagé, Peter (1991), *ENTIRE Function Server Technology Overview White Paper*, Software AG, June 1991

Chapter 15

ORACLE: Open CASE Environment

1 Introduction

In 1989 Oracle became, in terms of turnover, the world's second largest independent software company. Its growth, in little more than decade, has followed a typically product-led, entrepreneurial path: first came the dbms and forms-based tools, then the dictionary and method appeared, followed by a variety of end-user application templates, and finally the CASE toolset. All of these are targeted at multiple hardware platforms, to maximise their impact on the marketplace.

The Oracle product set, as shown in figure 15.1, has become centred around the dictionary product – CASE*Dictionary. Until recently the major draw-back has been the closed nature of the product, which made it difficult for third party products to integrate with it. The marketing message was that the user needed no products other than Oracle's.

Nevertheless, during the last few years, a number of third party CASE tool vendors have produced export/import exchanges with Oracle, mainly based on the generation of Oracle SQL schema statements. A market has also developed in providing better 4GL facilities than those provided by Oracle in SQL*Forms – for instance, products such as Uniface from SIA, Powerhouse from Cognos, Top Tools from ISS and Ally from UNiSYS.

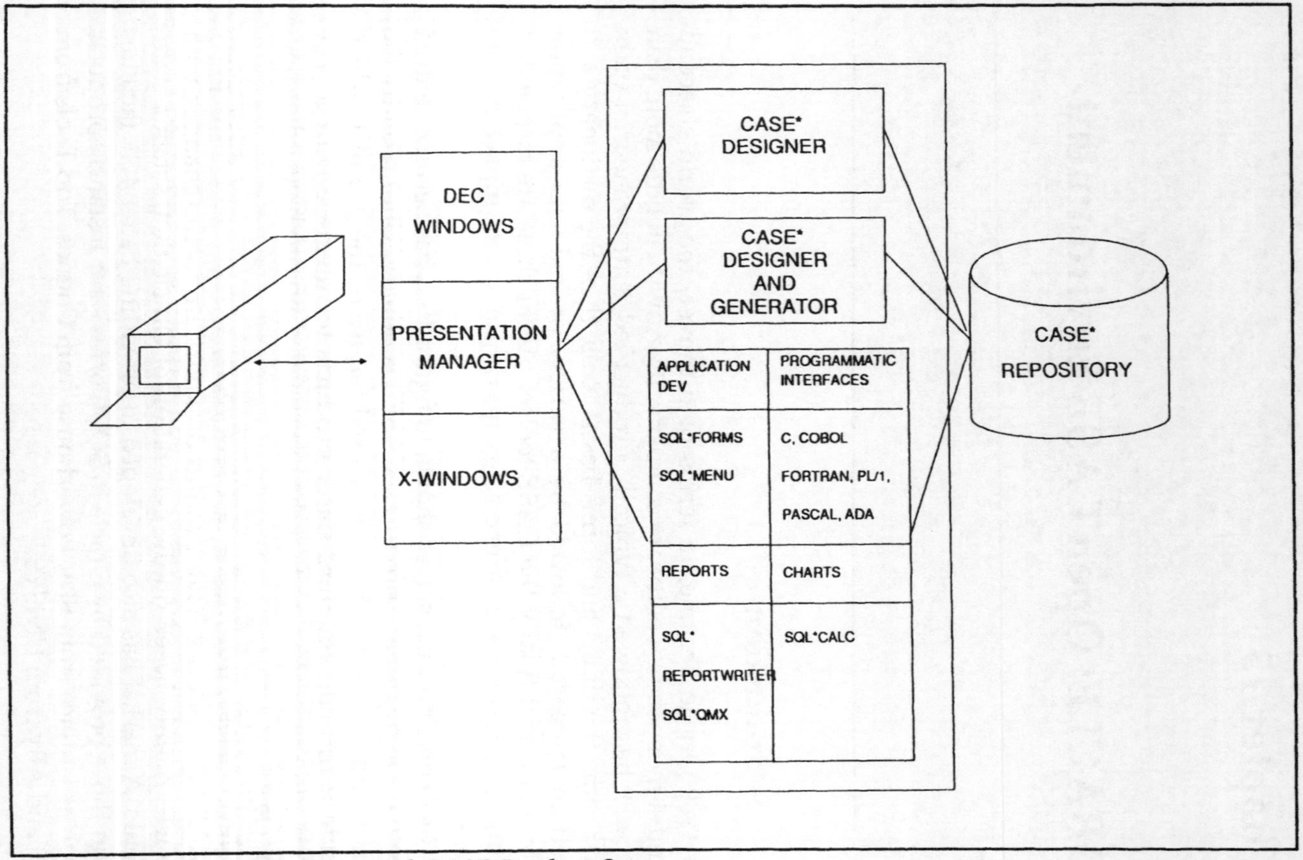

Figure 15.1 The Oracle Integrated CASE Product Set

At the beginning of 1990, Oracle bowed to the inevitable and announced the opening up of its CASE*Dictionary.

2 The purpose of the Open CASE environment

There are two basic ways to view the Open CASE environment. On the one hand, it is Oracle responding to IBM's AD/Cycle announcement, and demonstrating that it will provide a way of communicating with the repository manager (RM/MVS). On the other, it is the company's response to industry trends for more openness and for the seamless integration of software.

Oracle's stance is to insist that it has been at the forefront of these industry trends. With the experience gained from developing its own CASE products, the company has confidence in its in-depth understanding of open CASE environments and their development. Oracle has stated that, as CASE technology develops, creating additional requirements, so a significant percentage of its development effort will be targeted at broadening the openness of the Oracle 'CASE solution'.

3 Oracle's views on aspects of openness

The scope of Oracle's CASE solution, it believes, will be determined by a number of aspects of openness, in conjunction with the technologies and product characteristics necessary to implement them. The aspects identified are:

- portability
- network support
- method support
- repository extensibility
- tailorable graphics
- integration of components
- standards compliance
- system generation.

3.1 Portability

Any complete CASE environment must support the hardware and software infrastructure within the organization. This is critical to its acceptance and its ultimate success. Most companies have adopted a pragmatic, multi-vendor approach to hardware and software procurement, and wish to retain the freedom to do so. Applications are therefore developed for a range of different hardware and software platforms, with a growing trend towards cross-systems development.

Oracle addresses these heterogeneous computing environments, as its CASE solution was designed to be inherently portable. CASE*Dictionary's use of the Oracle relational database to hold the repository information ensures that minimal porting effort is required between platforms. This is aided by the provision of development tools such as SQL*Forms and SQL*Plus, and by the use of the C programming language. CASE*Designer runs under X-Windows, DECWindows and Presentation Manager, maximising the range of graphics-based hardware and software options. CASE*Generator for SQL*Forms, the first of a family of CASE*Generator products, produces applications ready for use on any of the hundred or so platforms supported by SQL*Forms. The resulting applications have automatic support for the different styles of terminals that can be used.

3.2 Network support

Another key aspect of the heterogeneous development environment is the network, providing the ability to combine different machines into a working system. With development staff using different machines, potentially at different locations, there is a strong requirement for the CASE environment to facilitate protocol independent network support. If this cannot be achieved then development can become fragmented, with isolated developers replicating effort or, perhaps worse, progressing in different development directions.

Oracle's approach to CASE utilizes the protocol independence of SQL*Net, its distributed database networking product set. This product ensures that the information held within a shared CASE*Dictionary is fully available to all network users, regardless of their location or preferred access style. The open, distributed use of

CASE tools across a mixed vendor, mixed protocol network allows users to exploit CASE technology to the full, regardless of the hardware and software infrastructure within their organization.

3.3 Method support

Numerous development approaches are used by today's organizations. Some IT departments have elected to use proprietary structured methods, such as Yourdon, SSADM or varieties of Information Engineering, whilst others have taken various techniques from structured methods and combined them with their own systems development experience to produce custom methods. This diversity of development style places a strong demand on the CASE environment: it must be sufficiently flexible to accommodate existing methods, and must adapt to changes as new techniques are adopted.

Oracle's CASE tools were originally based on the techniques deployed by the company's CASE*Method, an Information Engineering variant. The tools will continue to support this approach, and will offer additional facilities, over time, to increase the computer assistance offered to CASE*Method users. Support has also been added, and will continue to be added, for techniques that do not feature as major components of CASE*Method.

3.4 Repository extensibility

The repository component within a CASE environment acts as the database for the systems development activity. The repository records and cross checks all development-related information to provide a common, agreed and shared definition of work done and to be done.

Today, definitions held within CASE*Dictionary can be expressed using graphics, forms, reports and defined views across the repository structure. Users can access CASE*Dictionary views using any Oracle tool, and can tailor the output to their precise needs. Although vendors can attempt to provide support for all element sets within their repositories, users will always demand additional information. Facilities to extend the repository, to tailor it more precisely to users' needs, is a growing requirement for repository products. Various areas within the repository need extensibility, but the following are of key importance:

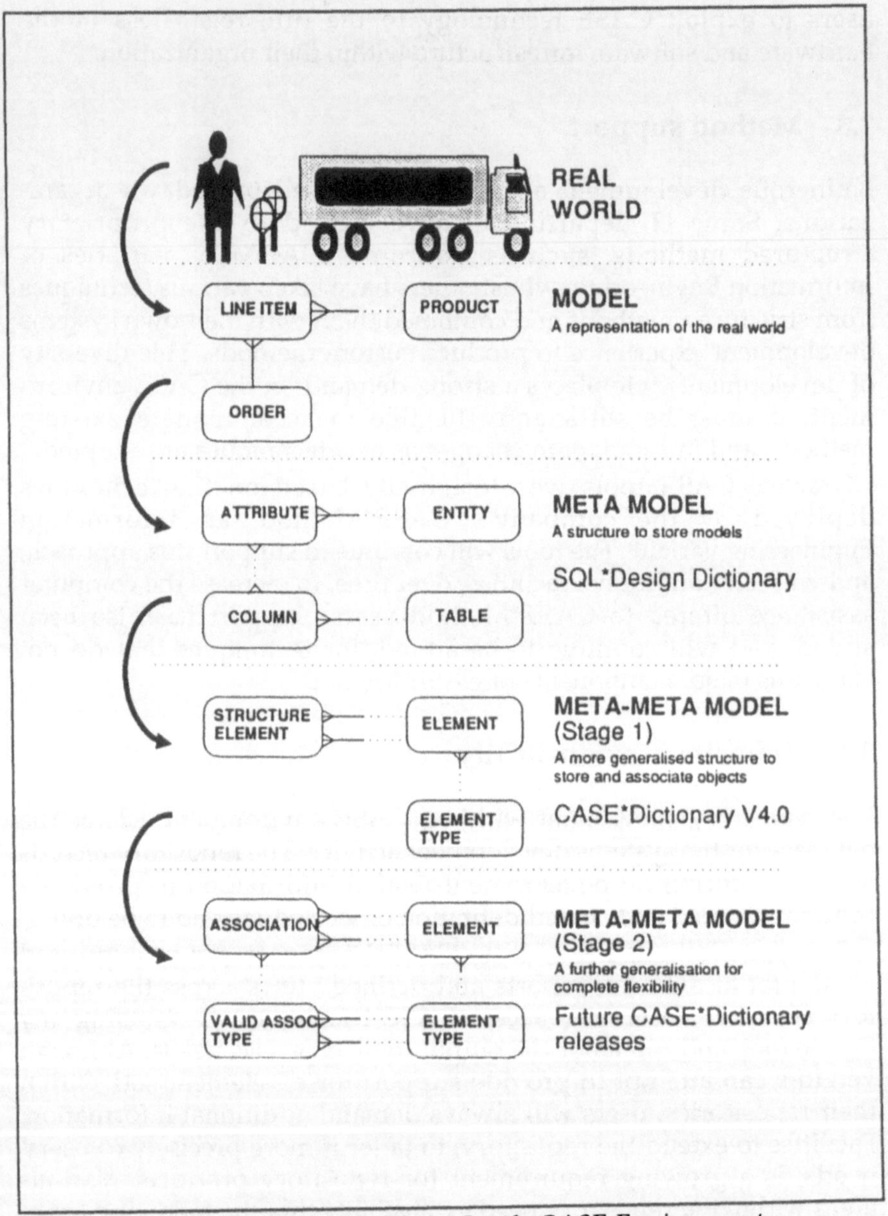

Figure 15.2 Abstract Data Models in Oracle CASE Environment

- it must be possible to specify new element types for the repository, in terms of name, properties and valid associations with other element types
- the tools that access the repository must have the ability to learn about users, and about changed element types and associations within the repository.

The architecture of Oracle's CASE products is evolving to support these requirements. Figure 15.2 illustrates this progression to a more abstract data model.

3.5 Tailorable graphics

Today, the functionality of CASE*Designer is partially determined by the structure and content of the underlying CASE*Dictionary. The matrix diagrammer determines which object types can be cross-related, by reading the object type and association definitions held within the repository. In future this will be less restrictive: using CASE*Dictionary version 4.2, user extensions are accessible via the matrix diagrammer.

Future versions of CASE*Designer will extend this capability to all diagramming facilities, allowing users to create new diagrammers, to manipulate new and existing object types and their associations, held within a fully extensible CASE*Dictionary. In combination with new graphical options for displaying objects, this should provide a fully tailorable and extensible CASE development environment which can be easily tuned to suit the exact requirements of the CASE users.

3.6 Integration of components

For a variety of practical and historical reasons, the CASE development environment within an organization may comprise software components from different vendors. Since the industry has not agreed on standards for the coexistence of multiple CASE tools, organizations must consider carefully how the components will fit together.

Through the use of ISO IRDS facilities, Oracle products will be interfaced to other dictionary products, starting with ICL's DDS. This connecting facility is known as CASE*Bridge. The products within this family are designed to move information, generally bi-directionally, between CASE*Dictionary and other repositories. The

ICL DDS connection will use an interchange file in ICL's DDCL format. Other CASE*Bridge products under consideration will use a more generalised mechanism. Oracle has identified a key rôle for IBM's repository manager (RM/MVS), and this will use the common programming interface of AD/Cycle. A more direct product-to-product interface is also to be produced. The first CASE*Connect product will be for IBM's AD/Cycle using the generic dictionary interface (GDI) in conjunction with IBM's common programmer interface (CPI). In the same announcement, Oracle stated its intention to conform to the ISO IRDS standards as these appear during the coming decade. Figure 15.3 shows this new set of open facilities.

3.7 Standards compliance

Multiple CASE tools could coexist easily if they all complied with a single, comprehensive standard. Sadly, suitable standards either do not exist or are not well supported. In the coming years the ISO IRDS standard is likely to be well supported but is not fully ratified today. Oracle is an active participant in ISO. The GDI (see section 3.6) accommodates the ISO IRDS concepts, and may be extended when IRDS standards are ratified; alternatively, Oracle may adopt a CASE*Bridge/CASE*Connect approach.

The lack of agreed standards is a significant obstacle to truly open CASE environments, but the architecture of the Oracle CASE tool set should allow relevant standards to be supported rapidly when they are ratified and adopted.

3.8 System generation

Effective system generation, producing 3GL or 4GL code with associated physical data storage definitions, can be based on user requirements and design decisions held in a repository. Target development environments vary greatly, owing to the plethora of programming languages, file and database systems and environments used across the industry. A truly open CASE solution would generate complete working systems in any development environment, regardless of the data storage technology or programming language used. Given the current state of CASE technology, this is impractical; however, a wider range of possibilities is emerging.

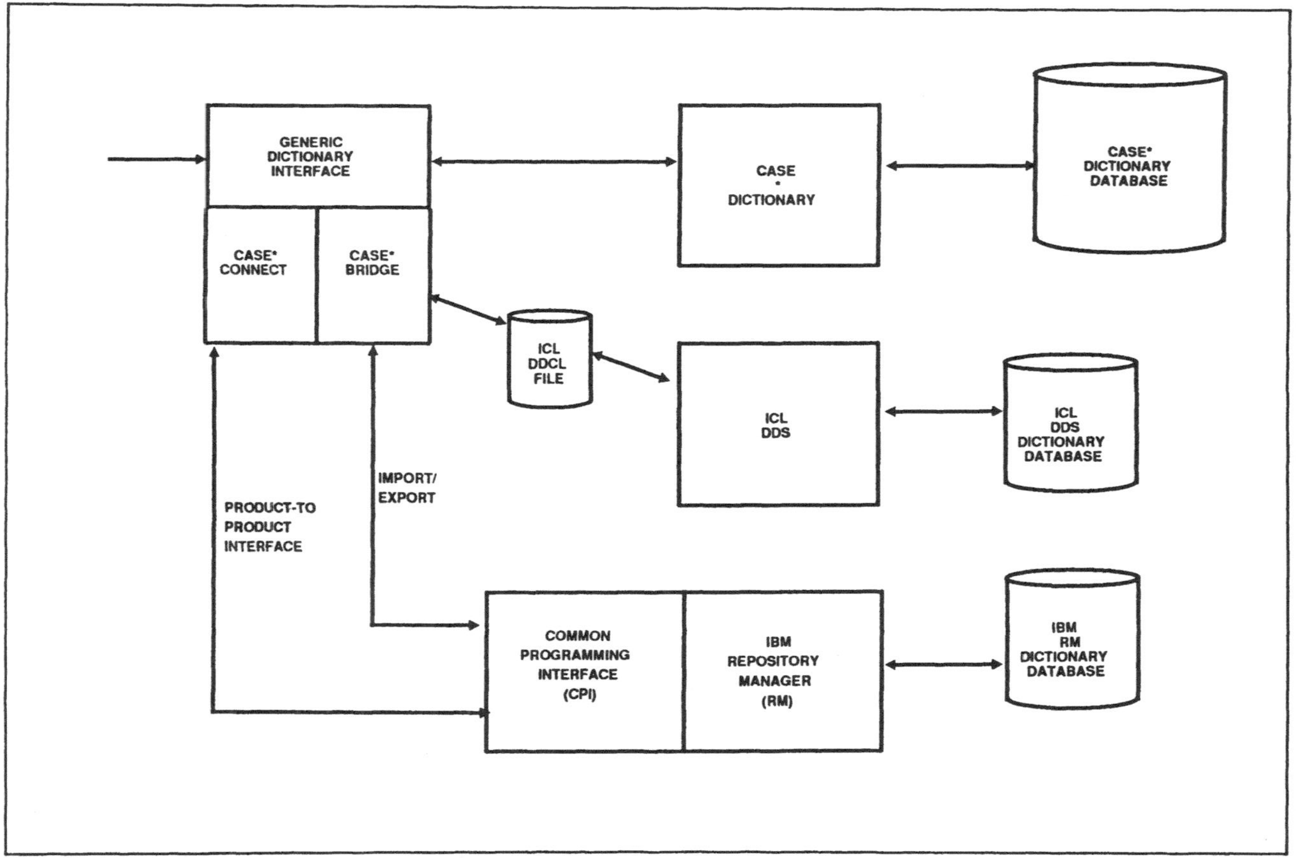

Figure 15.3 Oracle's Open CASE Product Set

Oracle's CASE solution delivers application generation for forms based programs using SQL*Forms, via CASE*Generator for SQL*Forms. Database definitions are created from CASE*Dictionary for both Oracle and DB2, using SQL*DDL. Future versions of CASE*Generator will support SQL*Plus, SQL*Report Writer and SQL*Menu. and longer term developments will focus on portable 3GLs with particular attention to C.

A real capability to generate 3GL code will have to wait for additional procedural specifications to be defined within CASE*Dictionary, but this will be implemented via a function logic language with an associated graphical interface through CASE*Designer in its next major release.

4 Open CASE environment summary

This strategy from Oracle represents a major change of direction. Its sales message was always that the company had the product set to 'carry you from business problem to business solution'. Oracle is now claiming that, for the last decade, it has been committed to supporting open computing. Some might beg to differ with this statement, but the current strategy does seem to provide a turning point. Indeed, Oracle's endorsement of the ISO IRDS standards was the first by a major US software company.

5 Further reading

1 Oracle (1990), *The Oracle Open CASE Environment Announcement*, February 1990

Chapter 16

MSP: RDADE

1 Introduction

Manager Software Products (MSP) was founded in 1966, and is best known for its dictionary product, DATAMANAGER, virtually a standard product in many IBM mainframe installations. MSP currently provides a complete family of programming products for the CASE environment – the MANAGER family – which provides an integrated information and systems engineering environment, designed to make the life cycle of planning, modelling, constructing, and operating information systems into a viable proposition.

The MANAGER family is driven from the corporate dictionary/repository framework which is provided by DATAMANAGER, DICTIONARYMANAGER and CONTROLMANAGER, and supports a variety of database environments – both hierarchical (IMS and DL/1) and relational (DB2 and SQL/DS). By providing standard and extensible, user-definable features, the MANAGER family supports organisations' information engineering needs, both current and future. Figure 16.1 shows the relationships between all the MSP MANAGER products.

2 MSP and IBM's AD/Cycle

MSP recognised that it was extremely vulnerable to AD/Cycle. In response to IBM's initial announcement in 1989, MSP chose their user

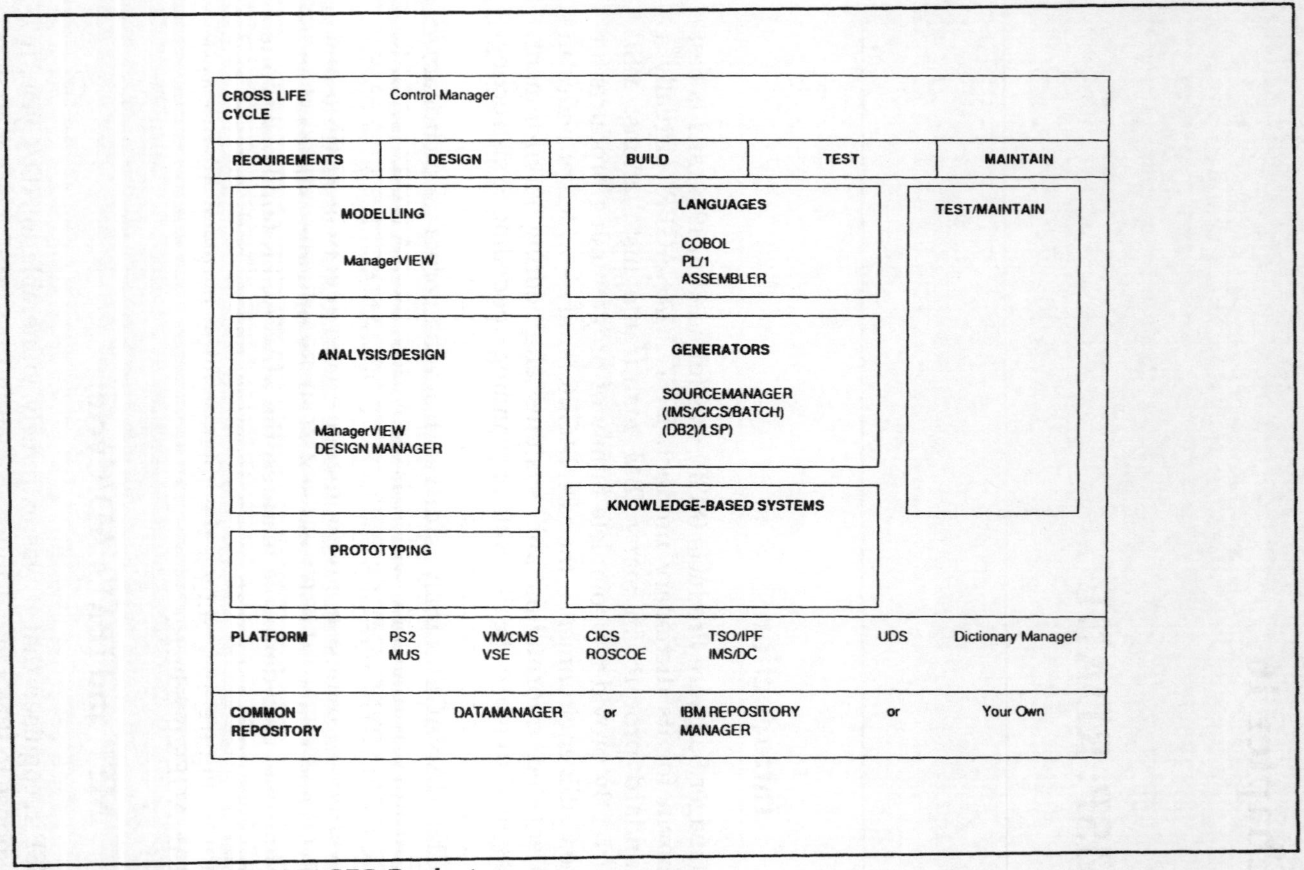

Figure 16.1 MSP MANAGER Products

group conference at the end of that year to propose three possible strategies for their users to follow:

- stay with MSP, and use RDADE to support your information management strategy (in effect – ignore AD/Cycle)
- co-exist, and choose when to use RDADE and when to use IBM's AD/Cycle products
- integrate, and allow MSP to provide an automated import/export facility between the two environments.

MSP produced their own paper and issued guidelines to their users on AD/Cycle. In these, they welcomed IBM's strategy on the grounds that:

- the announcements vindicated MSP's years of practical implementation of a corporate dictionary/repository as the foundation for an application development/CASE strategy
- the announcements endorsed and reinforced MSP's past, current and future strategic directions
- the announcements indicated that the AD/Cycle architecture would be open, and promoted the concept of integration between IBM and non-IBM offerings.

With AD/Cycle, IBM defined the concept of a synchronised environment for application development for the 1990s. MSP enables this concept (see figure 16.2) with METHODMANAGER. This automates not only the application development process but also the management of the corporate information resource. METHODMANAGER's open architecture means that a user can incorporate their choice of tools and methods into an integrated AD/Cycle strategy.

New capabilities within METHODMANAGER include toolset and life cycle services which deliver proven AD/CYCLE functions. These services fulfil the AD/Cycle objectives of:

- total life cycle management
- automation through the use of a life cycle model which:
  - drives the application development process
  - enforces automation by choosing the right tool.

MSP was one of the first vendors to receive AD/Cycle development information from IBM, and can translate and transfer information between the MSP and IBM repositories. IBM is required to release critical information about its repository to MSP (for example, the information model, and the CPI and API interfaces).

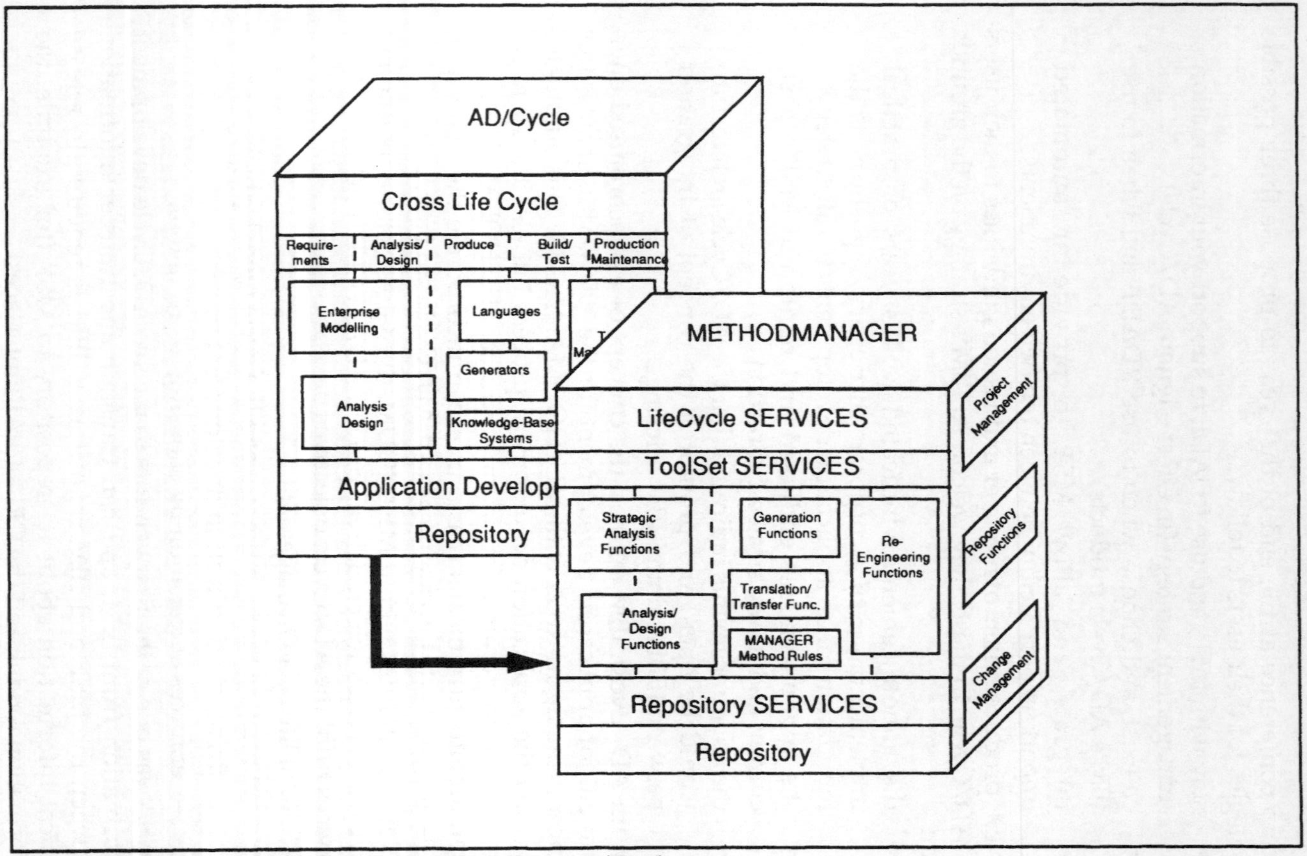

Figure 16.2 METHODMANAGER Enabling AD/Cycle

3 RDADE – repository driven application development environment

MSP's RDADE – Repository Driven Application Development Environment – strategy (see figure 16.3) is based upon the following:

- a new (or new in 1989) MSP product strategy
- conformance to IBM's AD/Cycle model and services
- the ability – where required – to co-exist with IBM's repository manager (RM/MVS)
- the adoption and exploitation of IBM's systems application architecture (SAA)
- the adoption of an open architecture approach
- the ability for users to tailor the MSP product line.

METHODMANAGER has been restructured to support the following objectives:

- automation of methods
- integration of tools
- management of the central repository
- adaptability to a company's needs and style.

Common user access is to be provided using a standard SAA CUA interface. On the programmer workstation this will include support for graphics, windows, mouse and pull down/pop up menus; on networked workstations additional support will be provided for menus/panels and in-context help.

Several methodology models are supported by METHOD-MANAGER, covering the following areas:

- life cycle
- activity/tool usage
- diagram schema (allowing diagram symbols to be amended)
- documentation
- export/import
- schema/repository information.

A set of life cycle services is provided to allow navigation through the life cycle setup and to manage projects. These services were test marketed in Germany. Three life cycle types are supported:

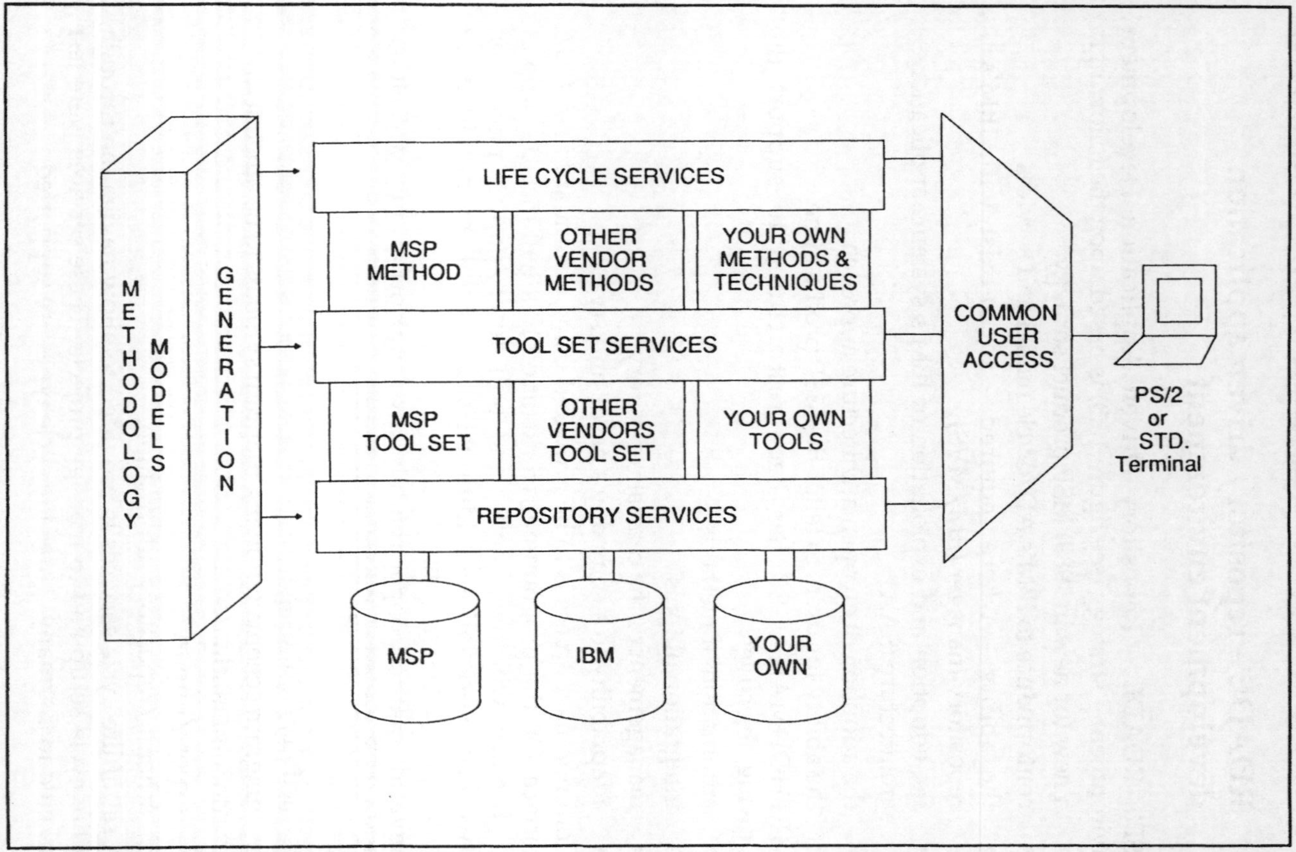

Figure 16.3 MSP's RDADE Strategy

- strategic information planning
- application redevelopment (of all sorts)
- data management.

Figure 16.4 shows the life cycle structure which is supported; this is based on the DATAMANAGER status facility.

RDADE users can choose to use their own tools, or select them from MSP or other vendors. A set of toolset services enables the chosen set to be integrated. Again, a German version of these services has been test marketed. MSP's own toolset consists of:

- diagramming/graphical editor
- end user reports/program
- matrix manipulation/affinity analysis
- dialogue design and screen/report printing
- application generation/prototyping
- logical and physical database design
- information resource management
- repository export/import reconciliation
- reverse engineering
- language support for COBOL, PL/1 and Assembler
- multiple DBMS support
- multiple IBM operating environment support.

A set of repository services, including configuration management capabilities, will provide standard access to any dictionary or repository.

4 MSP's own CASE product – managerVIEW

MSP's front end CASE offering is managerVIEW. It is an interesting tool, as it can run in three different modes depending on the user's requirements:

- stand-alone, where the user can access all PC-based managerVIEW features
- loose-host-connectivity, where the user specifies whether commands are for local or mainframe processing
- close-host-connectivity, where the product automatically routes all commands.

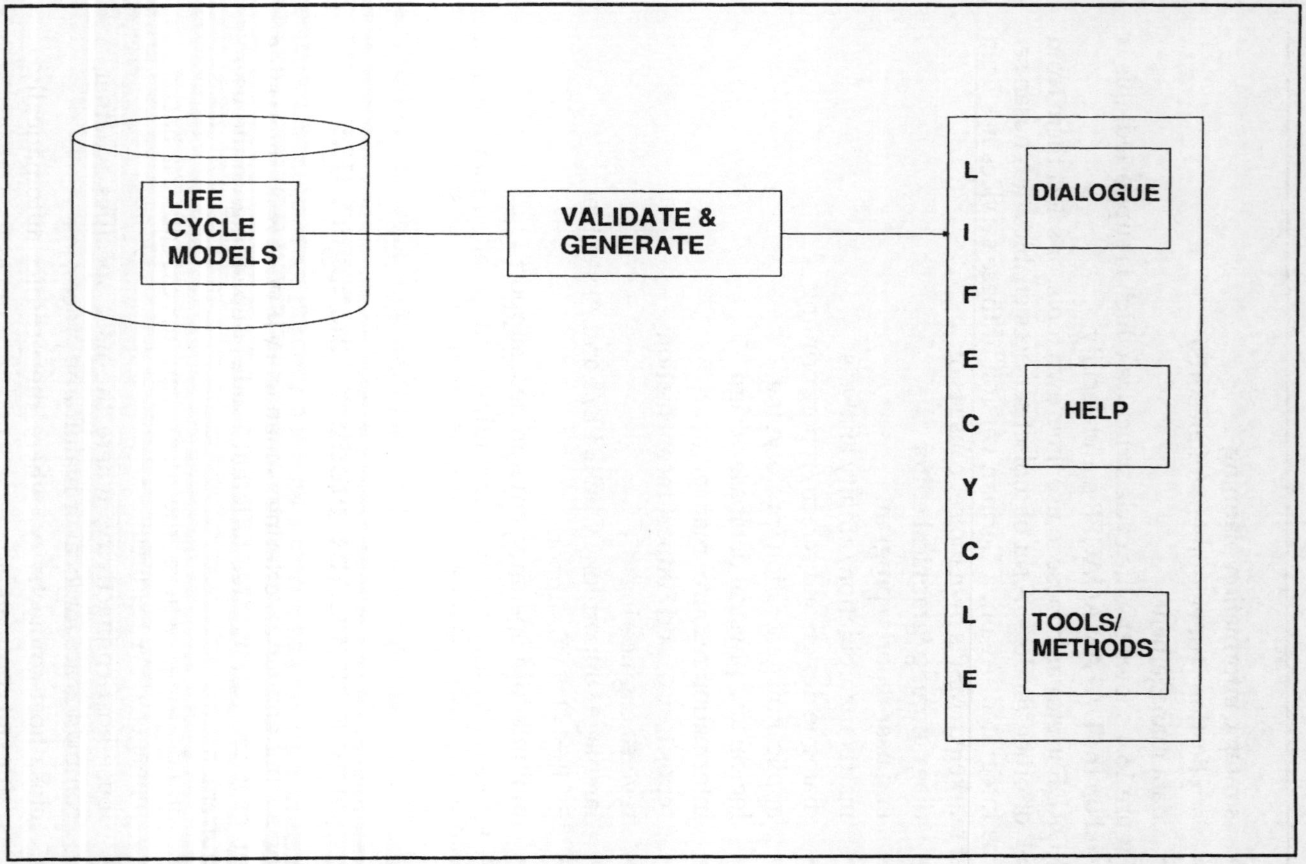

Figure 16.4 MSP's RDADE Life Cycle Structure

In addition, managerVIEW is tailorable, allowing the user to define diagram shapes, menus, validation rules and diagram types. There is a pre-tailored version of managerVIEW to support SSADM.

However, it is significant that MSP has not been among the leaders in making strategic product announcements over the last two or more years. The authors consider it a strong possibility that MSP, despite its avowed intention of maintaining its strong position in the dictionary/repository sector of the CASE industry, is in danger of slipping out of the race. The next year or so will tell.

5 Further reading

1 MSP (1990), *MSP and AD/Cycle*, MSP-AD

2 UKDUG (1989), *Guidelines to Manager Product Users regarding IBM's AD/Cycle announcement*, UKDUG 11, 16-18 October 1989

3 MSP (1989), *Relational Technical Support in RDADE*, MPR-RTL, July 1989

4 Hassel, Eric, (1989), *MSP Direction and the ADE*, UKDUG 11, Session 27, 16-18 October 1989

Chapter 17

Softlab: Maestro II and E-CASE

1 Introduction

This book includes a discussion on Softlab, not only because it is an important CASE vendor, but also because it is different. While the vendors described in other chapters are (mainly) American based, or certainly from an Anglophone background, Softlab is a German company. Unlike most of the others (apart from Texas Instruments – see Chapter 19) it is primarily a single-product vendor, but that single product is targeted at the entire application development arena, including full life cycle integration, management of the development infrastructure and multiple architecture targets. It is as if IBM had introduced AD/Cycle as a single software product, rather than as a framework into which to slot many products from numerous vendors.

As indicated in Chapter 1, Softlab's Maestro II product incorporates what Softlab's Chairman, Dr. Klaus Neugeberger,[1] identifies as the important features of an IPSE – tightly integrated life cycle tools for project and configuration management, with multiple architecture targets – and of I-CASE – front end and back end tool integration. 'For historical and cultural reasons, the I-CASE concepts will continue to thrive in the USA. The IPSE concepts will continue to do well in Europe. Those users and vendors who can incorporate the two the quickest will benefit early from the emerging E-CASE movement of the 1990s.'

I-CASE	IPSE
TIGHT INTEGRATION OF FRONT END TOOLS WITH SPECIFIC BACK END GENERATOR	OPEN TO DIFFERENT TOOLSETS, PARTICULARLY AT THE BACK END, BUT NOT TIGHTLY INTEGRATED
SINGLE TARGET ENVIRONMENT: NORMALLY IBM, DIGITAL OR HP	HETEROGENEOUS TARGET ENVIRONMENTS, EVEN FOR SINGLE SYSTEMS
UNIX SUPPORT LACKING	UNIX-BASED
STANDARD DEVELOPMENT METHODOLOGY: MOSTLY INFORMATION ENGINEERING	METHODOLOGY INDEPENDENT
CUSTOMISATION NOT USUALLY POSSIBLE	FULLY CUSTOMISABLE, WITHIN THE LIMITS OF AVAILABLE TECHNOLOGY
LACK OF INFRASTRUCTURE OR LIFE CYCLE SUPPORT FUNCTIONS	CONCENTRATION ON INFRASTRUCTURE OR LIFE CYCLE SUPPORT FUNCTIONS: – PROJECT MANAGEMENT – CONFIGURATION MANAGEMENT – INTRA-TEAM COMMUNICATIONS – PROGRAMMER-ORIENTED TEXT SYSTEMS – MIS OFFICE FUNCTIONS
COMMONLY SINGLE USER	MULTI-USER
IMPRESSIVE GRAPHICS CAPABILITIES	GRAPHICS POOR, OR SUPPLIED BY INTERFACED PRODUCTS

Figure 17.1 IPSE and I-CASE – Typical Characteristics

Figure 17.1 distinguishes between the 'traditional' I-CASE and IPSE philosophies. The E-CASE philosophy is to bridge the gaps, combining the undoubted strengths of the two approaches.

It should be noted at this point that Softlab is not the only vendor breaking down these barriers. For instance, as discussed in Chapter 19, the classic I-CASE tool – Texas Instruments' IEF (Information Engineering Facility) – has moved rapidly of late and will, in the next couple of years, provide many of the E-CASE facilities which Maestro II already exhibits.

2 Background to the E-CASE concept

The evolution of software engineering can be broken, suggests Neugeberger, into six phases (see figure 17.2):

- in the late '70s, dumb terminals were linked to mainframe back end tools – code generators, compilers and testing facilities
- in the early '80s, stand-alone analysis and design CASE tools were introduced, with no automated connection to target machines
- by the mid '80s, interfaces had been developed between front and back end CASE tools, such as the Excelerator–Telon link
- in the late '80s, integrated CASE (I-CASE) emerged, with tightly integrated front and back end tools developed, sold and supported from the same vendor
- in parallel with the previous three phases, the '80s saw the European IPSE movement, emphasising tightly integrated life cycle tools such as project management and configuration management
- in the early '90s, E-CASE is bringing together the IPSE and I-CASE movements.

Softlab's background is as an IPSE vendor. Its Maestro I product, which Maestro II will ultimately replace, was the only IPSE to be relatively successful in the worldwide mass market, and Softlab installed 25,000 workstations targeting IBM, Bull, Siemens, ICL, HP, Digital, NCR and UNiSYS environments.

The largest US customer is Boeing, with over 300 workstations installed. What held Maestro I back from further success was its continental European origin, regarded as exotic for a software product in the USA, its relatively poor methodology support and (for the vital North American market) its UNIX platform. The cultural issues should not be underestimated either: Maestro I was about control, communication and standardisation. The rest of the world was perhaps more interested in methodologies and graphics.

Maestro I's continued use in large European organisations, such as the UK Inland Revenue, the Automobile Association (AA) and many public utilities, owes much to its capacity to facilitate communication. Maestro I can hold, and make available at any time, all of an organisation's standard documentation; nobody every needs to use an

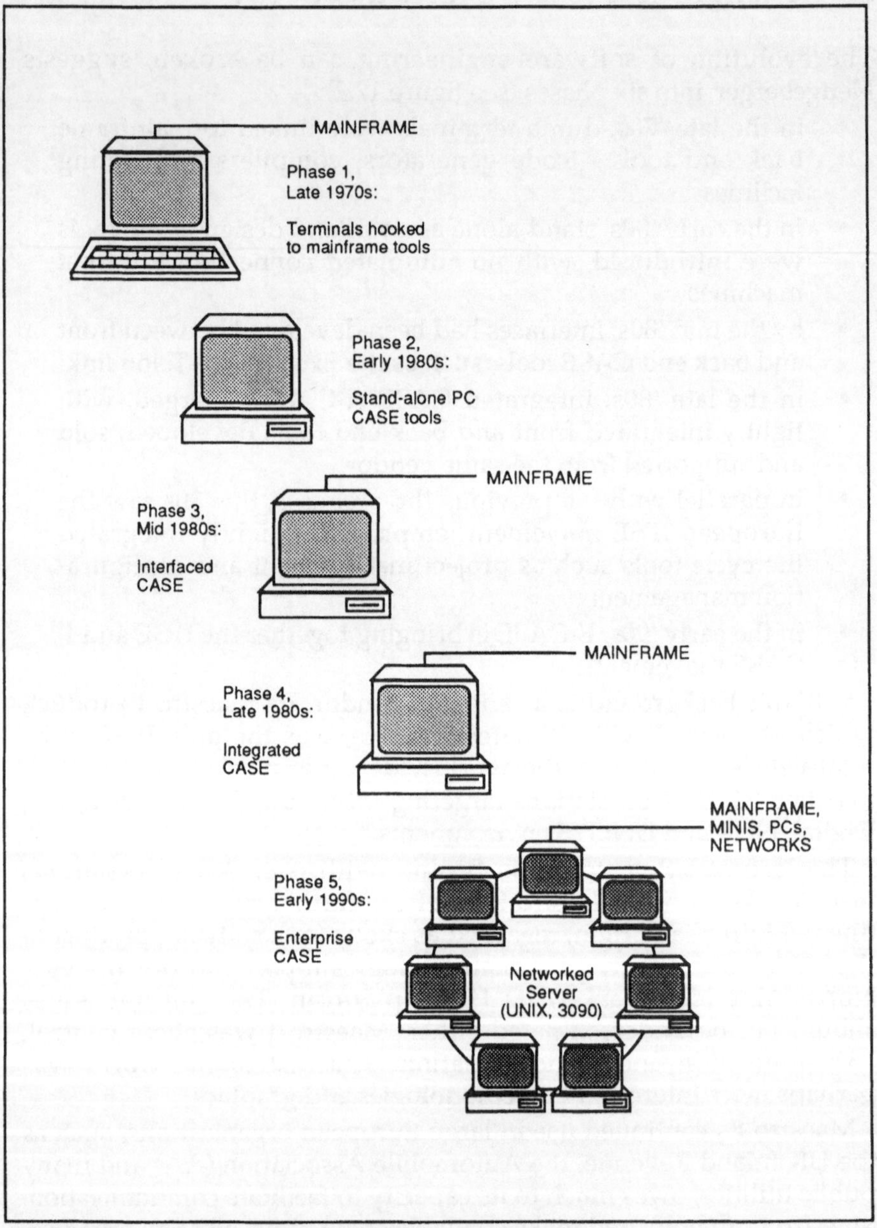

Figure 17.2 The Evolution of Software Engineering

outdated version and everybody can be told when something has changed. Project deliverables are made available simultaneously to all appropriate users.

The evolution to E-CASE can be seen as a natural one: given the three axes of CASE architectures (see figure 17.3) identified by Softlab:

- front and back end tool integration
- infrastructure/life cycle
- multiple architecture targets.

I-CASE products support the first of these and IPSEs the second and third. Sooner or later, someone was bound to bring them together in one product set. The IBM approach is AD/Cycle, discussed in Chapter 5, which allows (to use the Softlab terminology) IPSE and I-CASE functions to merge into the same software environment. IBM began talking about serving development needs at the enterprise level. Components include the following:

- infrastructure, such as the SAA information model, the DB2-based mainframe repository, the ADPS process model, the external source format for data sharing and other life cycle functions
- I-CASE, comprising Excelerator or ADW, integrated via the IBM repository with IBM's own CSP application generator
- multi-target IPSE, including MVS, VM, OS/400 and OS/2.

The major problem for IBM, of course, is the enduring and embarrassing lack of a mainframe repository with all the necessary facilities. This, combined with IBM's apparent internal differences, has tended to undermine the AD/Cycle programme, and to allow third party vendors – such as Softlab – to seize the initiative.

3 Structure of Maestro II

Maestro II is based around its repository, OMS, the object management system. This is an on-line, object-oriented database, with an accompanying methods data model, which together manage objects, attributes and relations. All tools and method functions reside at the client level while all data is stored at the server. See figure 17.4 which shows the overall structure of Maestro II.

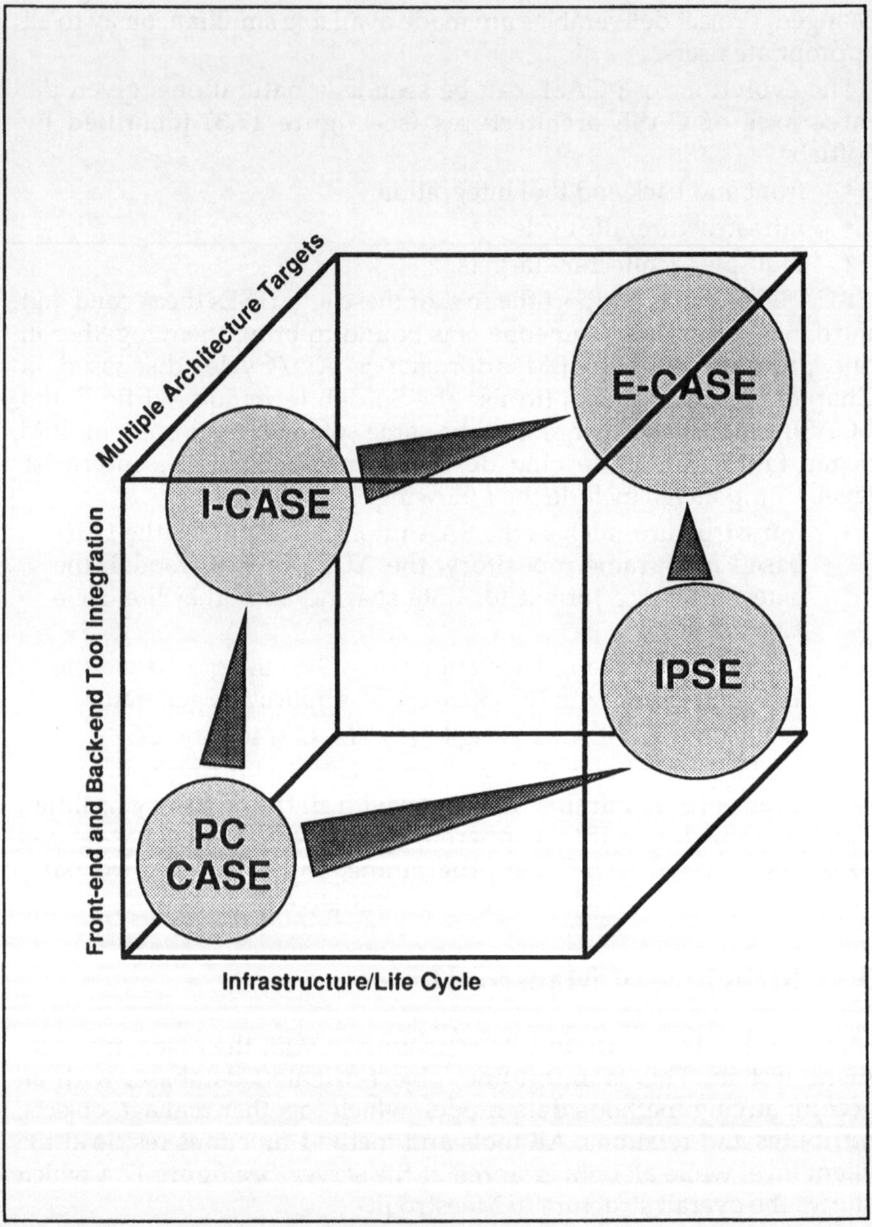

Figure 17.3 Evolution of CASE Architectures

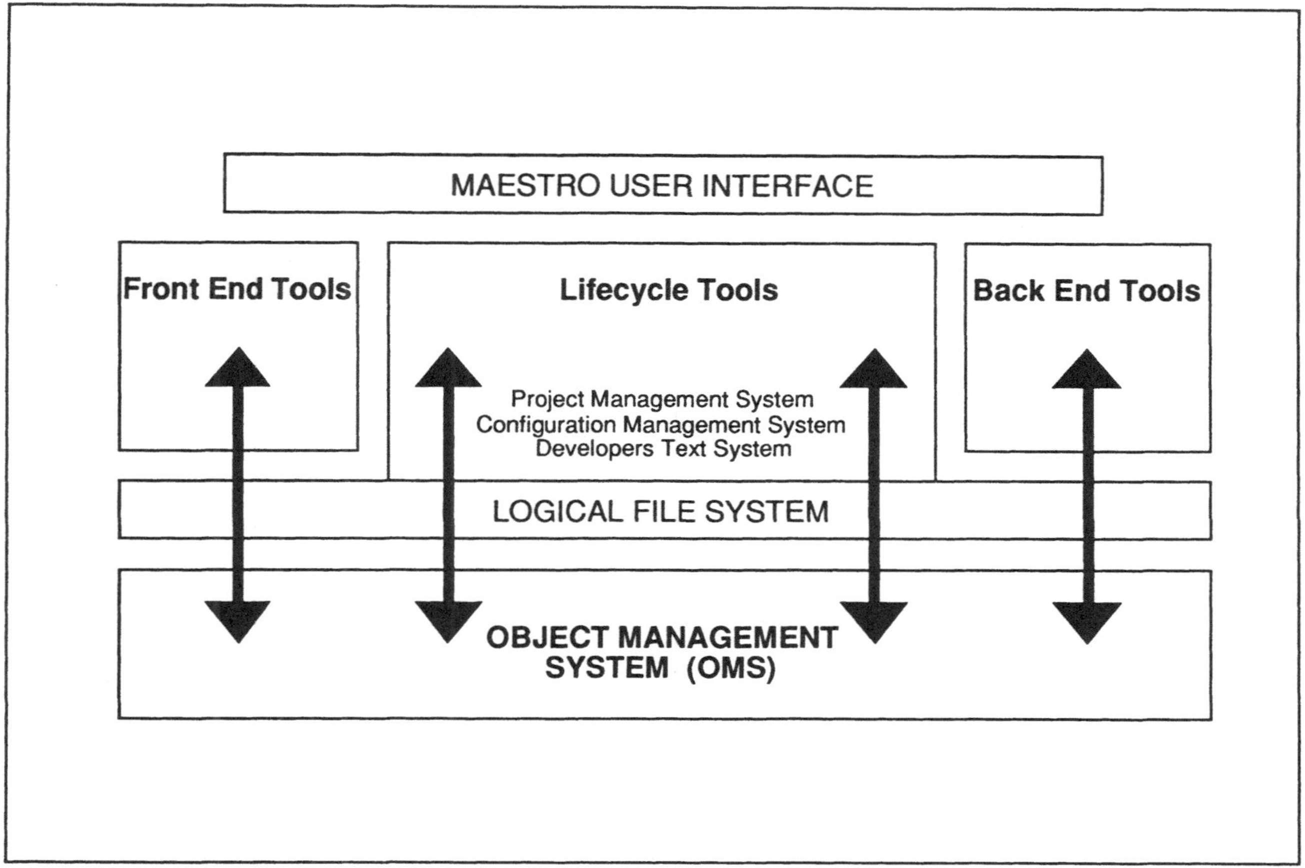

Figure 17.4 Maestro II Overall Structure

OMS acts as the repository for a suite of life cycle tools, such as:

- project management
- version control
- configuration management
- office communications
- a proprietary, high performance multi-user, multi-tasking windowing interface, due for CUA compliance in 1992.

The OMS meta-model and interfaces are published, so that customers can populate the OMS and the project management modules with their own data models and task models. The OMS is positioned as an on-line project database, and was designed to share information with the IBM repository.

At the front end, Maestro II offers a tightly integrated front end multi-methodology package, supporting various approaches, including SSADM, Merise, Structured Analysis & Design (SA/SD), besides being customisable to other methodologies. Softlab has recognised that for worldwide success, especially in the USA, it must support Information Engineering, on which it is actively working, and reverse engineering, where Maestro II's tool-building facilities will include applications for the take on of existing systems.

At the back end, Maestro II's knowledge-based code generator will store and reuse blocks of information for multiple-language and multiple-target environments. The first code generation facilities are scheduled for testing in the UK in 1992, in the form of a standard SAA CICS COBOL generator. Facilities will include automated testing.

A useful Maestro II feature is its scanners, which aid productivity by demonstrating instantly the effect of changes to screen layout parameters, or by loading old COBOL code for investigation, alteration and impact analysis. Effectively, there is an underlying reverse engineering metamodel. Existing code can be analysed and 'neatened', but it cannot currently turn spaghetti code into a structured program.

Different protocols – IBM, Digital, ICL – can be run as windows, and the contents cut-and-pasted between them.

Given a set of definitions of natural language constructs, Maestro II can be used to extract definitions and to define business rules, along the following lines: 'Peter's age is 5 years' can be analysed into 'Peter, age = 5 (years)'.

In addition, third party tools can be incorporated into the environment via its published interfaces. Softlab has signed marketing agreements, as global partners, with IBM (as a second-tier AD/Cycle vendor – see Chapter 5 – and as part of the AIX-CASE initiative – see Chapter 6 – and HP, and is involved with what it describes as regional players, such as ICL in the UK and Siemens in Germany. Local alliance are regarded as particularly important – what is right for the USA is not necessarily the answer in the UK, or in Germany.

The relationship with IBM Germany is a case in point. A recent article in *Infomatics*[2] explored the agreement whereby IBM Germany has agreed to market Maestro II once the porting to OS/2 is completed, and links provided between OMS and RM/MVS. The article expressed the significance of the move:

> The deal ... puts Softlab in pole position as the Case vendor most likely to provide the Lan-based Case tools to those IBM customers who want to develop in a distributed environment without having to step outside of IBM's AD/Cycle framework. More importantly, though, the agreement is the first instance of one of IBM's newly devolved national subsidiaries breaking ranks with IBM's development chiefs in the US, and seeking to accelerate the delivery of AD/Cycle functionality through a local alliance.

Softlab is not involved in ICL's OPEN*framework* (see Chapter 8) although it is an ICL CASE partner. The relationship is driven by large accounts and sales opportunities for DRS 6000s, rather than by collaboration on software, though ICL is regarded as a particularly co-operative partner.

Apart from those organisations already mentioned, Softlab regards itself as having a degree of synergy with the following:

- CCTA, over SSADM
- Software One, with its use of the latter's Exchange product to develop a link to ICL's DDS 850 data dictionary
- Hoskyns, over PMW
- MicroFocus, over COBOL 2 Workbench
- Object Management Group
- CA, over Telon
- Ingres Ask, over Ingres.

There is, however, a degree of potential conflict with the latter two, especially over future product directions.

Maestro II has certain limitations. For instance, it is not positioned as a complete presentation tool. It produces high quality output within a software development context – limited on fonts but excellent on (e.g.) tables – but relies on other products such as Microsoft Word for ultimate output quality. Versions created on such third party products can be managed under Maestro's configuration management system. Also, it cannot control the multiple scheduling of a resource, and Softlab suggests the use of PMW for this purpose. It can handle versions of attribute files but not, at the time of writing, variants.

4 Maestro II technical issues

The information in this section is based on Softlab's Migration Planning Guide[3] for existing (Maestro I) customers.

4.1 Components

The Maestro II base system provides the software platform upon which all the other components run. It consists of:

- user interface and desktop, intended to achieve CUA compliance in 1992
- developers' test system (DTS), a text processing system capable of handling files up to 132 bytes width
- electronic mail
- printing, including full Postscript support
- distributed file system (distribution is only possible at the file level, not at the object or attribute level)
- object management system (OMS)
- Maestro customising interface (MCI)
- tool customising interface (TCI)
- graphics editor (GED)
- browser
- administration, supplementing the standard UNIX functions.

In addition, Maestro I Link provides full communication with the Maestro I environment.

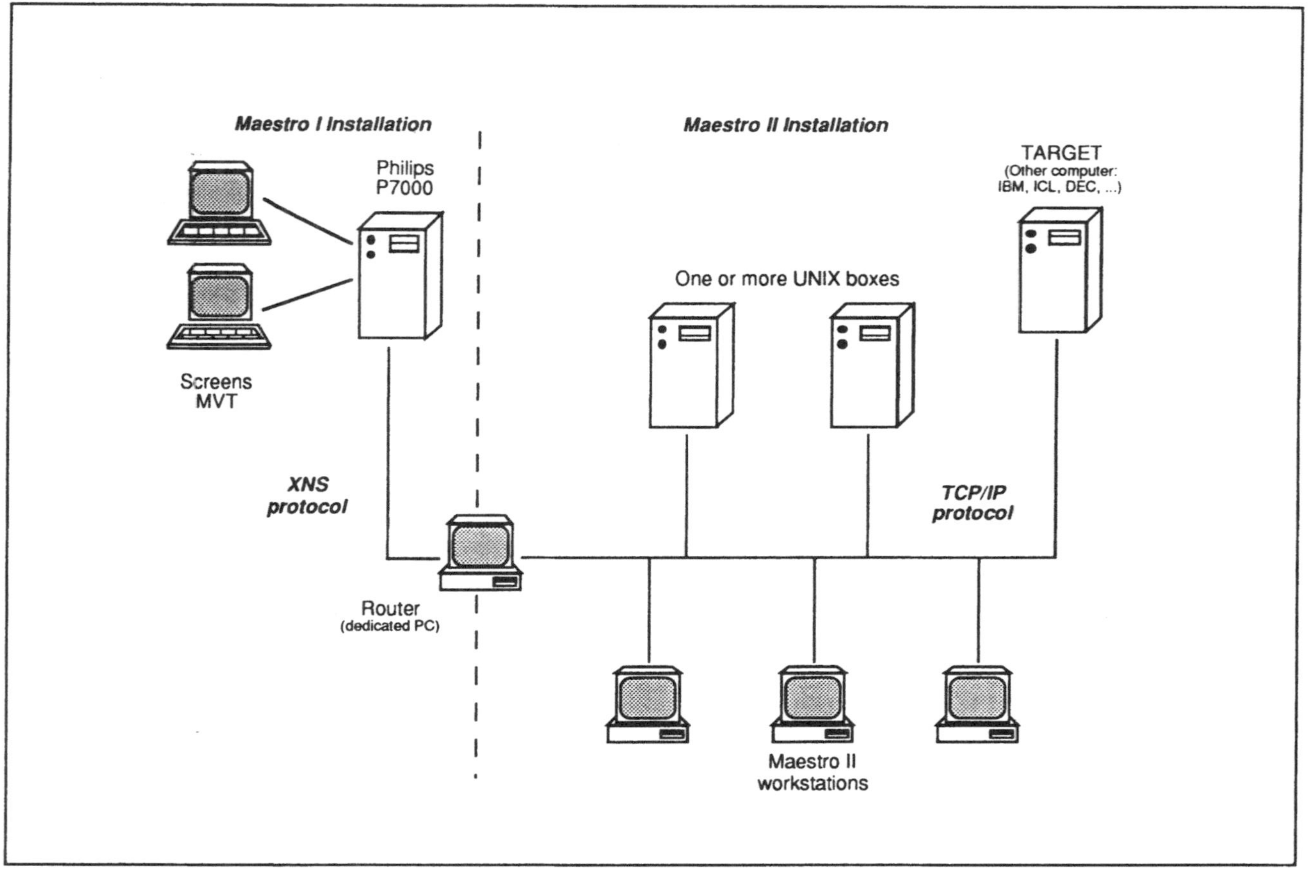

Figure 17.5 Typical Maestro Installation

4.2 Configuration

In a typical Maestro configuration, containing one or more Maestro I systems and Maestro II workstations and/or servers, two different Ethernet protocols are run:

- Maestro I (on Philips P7000 UNIX boxes with graphics workstations) uses the IDP component of the XNS protocol
- Maestro II uses the UDP component of the TCP/IP protocol.

A LAN router, consisting of a PC with one Ethernet card, is required to convert between IDP and UDP. Cross-system file access is then allowed. Figure 17.5 shows a typical installation, where Maestro I and II coexist on the same network, and where projects are progressively migrated to the new environment. The ultimate environment is simpler, as the Maestro I elements are removed from the system.

Maestro II workstations can be attached to Ethernet (IEEE 802.3) or Token Ring (IEEE 802.5) local area networks. LAN cards are supported for he following environments:

- Ethernet, PC/AT or EISA bus
- Ethernet, IBM microchannel
- Token Ring, PC/AT or EISA bus
- Token Ring, IBM microchannel.

4.3 Workstations

Maestro II will run on any PC or PS/2 with 2Mb. of memory and at least a 286 chip, though a 386 is recommended for performance reasons. A mouse and EGA or VGA graphics are required.

4.4 Servers

The OMS is released for a range of UNIX platforms in conjunction with the new workstation software. The software has so far been ported onto the following platforms:

- IBM – RS/6000
- ICL – DRS/6000
- Digital – RISC System
- HP – 9000
- Philips – P90X0.

4.5 Communications

Communication facilities are provided for IBM, ICL and, to a limited degree, Digital.

The IBM facilities are based on the 3270 file transfer facility (LU Type 2), and can be supplemented by 3270 RJE support, implemented using third party products. New facilities are provided at the workstation for control of the 3270 emulation software on the server via a batch link interface.

ICL file transfer facilities are implemented on a PC server on the LAN. Queues for services are managed on a designated Maestro II server and channelled via the PC server. Additionally, batch run profiles are constructed and submitted through ICL FTF to send and receive files.

The Digital support is limited to VT 220, and the target environment must support TCP/IP and Telnet.

5 Maestro II in use

Maestro II is aimed at supporting large development teams, running to hundreds of users. Softlab does not believe that it is really suitable for customers with less than 50 to 100 users. Target markets are based on vertical sectors – finance, government and utilities – and on target platforms, such as those where ICL and IBM mainframes are the primary architecture.

Methodology support tends to be country-specific; in the UK it is targeted at SSADM Version 4 sites, in France at Merise/2 and in Germany at SETec, but the product is inherently extensible and customisable. PCMS, the Maestro II project configuration management system, contains tasks lists for the various methodologies.

Figure 17.6 shows the relationships between the OMS and a given client's users, servers, projects, tasks, deliverables and systems.

6 Maestro II, E-CASE and the future

In the absence of a real, functioning AD/Cycle environment, Maestro II can only go from strength to strength in the large installations commonly found IBM world. The stumbling block, as ever, is the

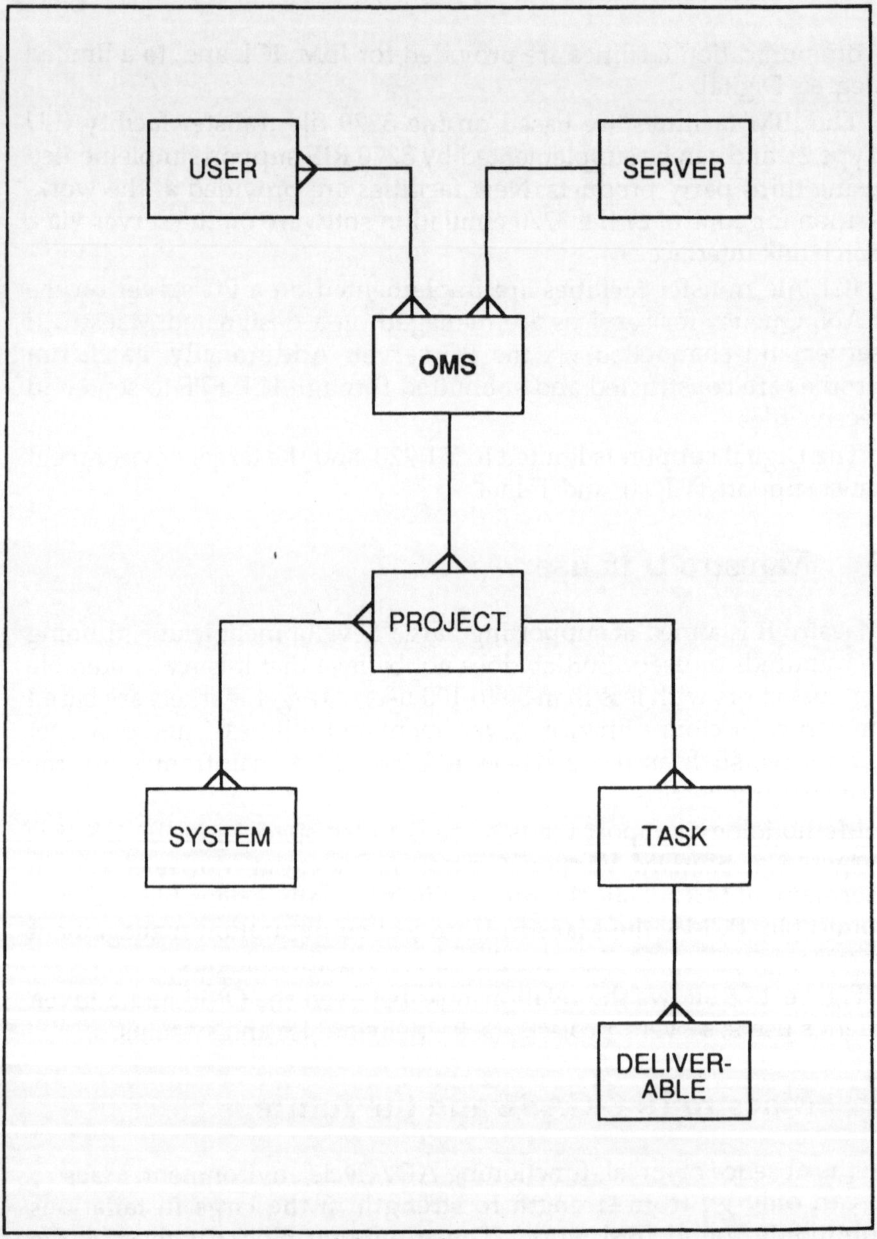

Figure 17.6 Customer Use of the OMS

North American market. While the UK has, by-and-large, accepted the principles of AD/Cycle yet shopped elsewhere for solutions, the USA has actually tried to implement it, favouring those vendors and products which possess the IBM 'seal of approval'.

If Softlab can penetrate the largest market in the world, and some observers have seen this occurring via an outright purchase of one of the IBM 'inner circle' such as KnowledgeWare or Intersolv,[2] then Maestro II may be in a position to challenge Texas Instruments and others for domination of the large-scale development environment of the near future.

7 References

1 Neugeberger, Dr. Klaus (1990), *The International Emergence of Enterprise CASE*, article in 'Peopleware', Computer Task Group, Inc., December 1990

2 *Infomatics* (1992), *German axis to target AD/Cycle*, April 1992

3 Softlab (1991), *Maestro II Migration Planning Guide for Maestro I Customers*, July 1991

4 Softlab (1989), *Maestro II*, leaflet

Chapter 18

LBMS: Engineer Framework

1 Introduction

LBMS was established in 1977 by a breakaway from BIS Applied Systems. The company, originally specialising in 'Methods for Information Technology', now has offices not only in the United Kingdom, but also in the USA. They first released a front end CASE tool in 1985 – Automate, which later became Automate Plus – and this has now sold over 6,000 copies worldwide. In 1987, it became clear to LBMS that the software industry was undergoing a major upheaval, as a second generation of PC software was superseding the first. This new generation incorporated two new, key technologies:

- graphical user interfaces (GUI), and
- local area networks.

From the feedback received from their Automate users, LBMS identified certain opportunities presented by the new technology:

- multi-user operations to support team working and the sharing of information,
- industry-standard SQL-based client-server architectures to help speed up access to information and provide better levels of reliability, and
- software (in particular Microsoft Windows and IBM Presentation Manager) that enabled multi-window, multi-tasking operations based on a standard graphical user interface (via IBM's CUA).

LBMS decided on a complete replacement of its CASE software to take advantage of this new technology, and started to develop a new front end CASE tool, Systems Engineer, which was released in February 1990.

While development of Systems Engineer was under way, LBMS identified another major trend in the marketplace, namely Integrated CASE or I-CASE. Realising that this concept was becoming the standard against which CASE products were being judged, LBMS decided that they had to be prepared to participate in this emerging market. To enable them to do this they pursued a strategy based on acquisitions and strategic partnerships to assemble an integrated tool set. The key organisations that LBMS acquired were:

- Meta Systems of Ann Arbor, Michigan, who had developed and sold a portable repository used by some 300 organisations, many of them in the US defence and academic sectors
- Michael Jackson Systems Ltd., who had developed a code generation tool as well as having experience in the real-time systems arena
- Keith London Associates Group, with their expertise in reverse engineering, as well as industry recognition in training.

LBMS has formed partnerships with IBM, with whom they are a member of the Tool Vendor Assistance Program (part of AD/Cycle), with ICL and with UNiSYS. The result of these recent changes has been, in effect, to transform a methods vendor with supporting CASE products to a serious software vendor also able to provide integrated methods support.

2 LBMS' definition of I-CASE

LBMS has identified the following elements that provide a common understanding of the definition of I-CASE:

- an integrated set of software tools, operating within a common framework, supporting all aspects of systems development, maintenance, reverse engineering and re-engineering, that interface to one another, inter-operate based on a common definition of metadata, and transfer information between themselves without loss of meaning

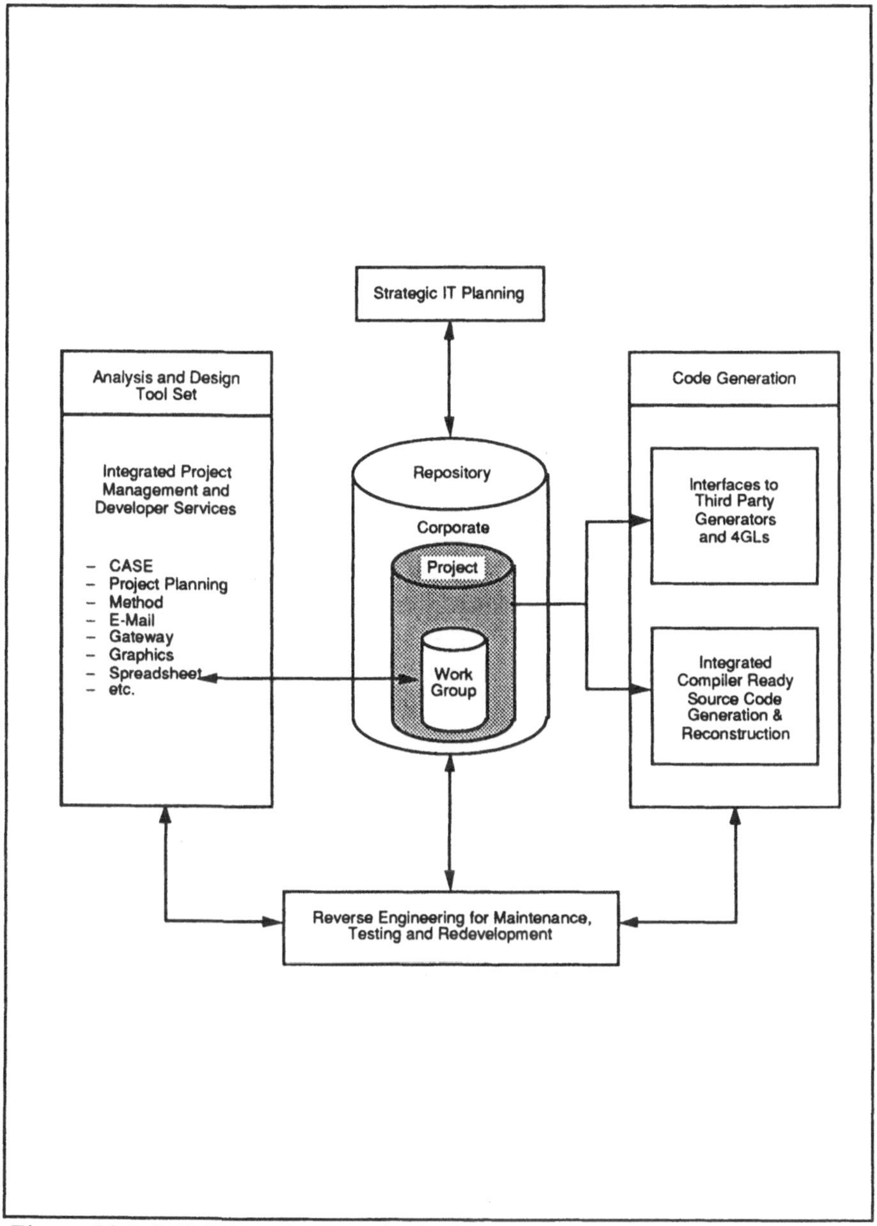

Figure 18.1 LBMS' Definition of I-CASE

- a common repository of information, in which the common metadata may be held, controlled, shared and easily accessed and manipulated
- the ability to generate 100% of the code required for an application, including definitions of any database objects required to support processing of the application
- full system development life cycle support from strategic planning through analysis, design, construction and implementation, and including reverse and re-engineering
- support for most of the activities undertaken by a developer including close integration with related tools in the graphical environment.

3 LBMS' I-CASE product set

The I-CASE product set consists of the following, arranged in LBMS' conceptual framework used to define I-CASE:

- Systems Engineer – a front end CASE tool for Microsoft Windows and OS/2 Presentation Manager, based on an SQL client-server database architecture
- Jackson Workbench – a UNIX-based front end CASE tool for Jackson, based on the IPSYS meta-CASE product
- Information Manager – a central repository running under OS/2 Presentation Manager
- SE/Construction (also known as Applications Engineer) – a code generator for the IBM environment
- SE/OPEN – an interface product based on CDIF draft standards that allows Systems Engineer to be integrated with many leading back end CASE tools
- REVENG – a set of reverse engineering and re-engineering facilities that are closely integrated with Information Manager
- LBMS Workplace – a graphical working environment built through a combination of Systems Engineer with dynamic integration with Windows-based desktop products, including word processing, presentation graphics and project management

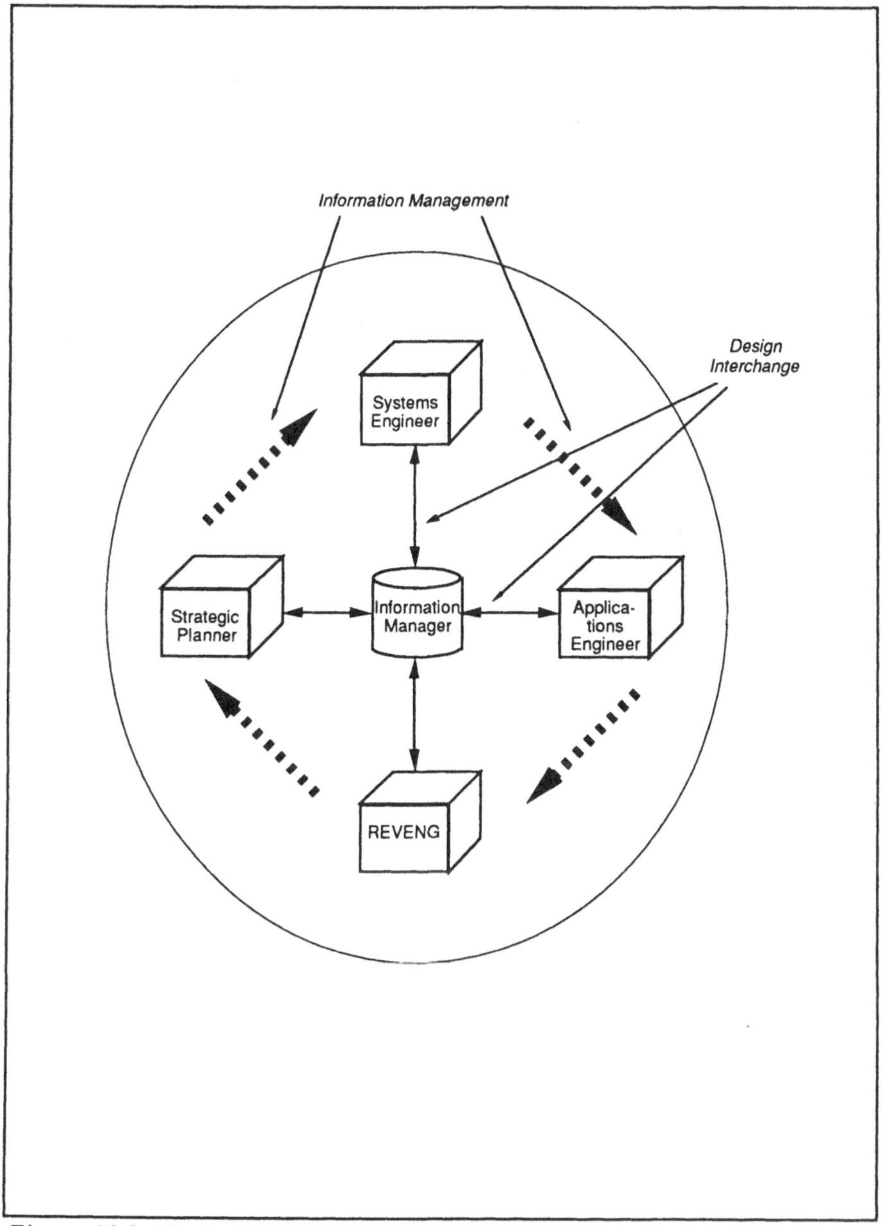

Figure 18.2 The LBMS I-CASE Product Range

- Project Engineer – a front end to Microsoft Project for Windows, which supports estimating and work breakdown structures in close integration with Systems Engineer
- SE/GUI – a generator for graphical user interface applications including Windows and Presentation Manager running C and Presentation Manager running COBOL
- Methods Engineer – a hypertext on-line help facility for methodologies, currently supporting Systems Engineering (formerly known as LSDM) and SSADM.

4 Systems Engineer: the front end CASE tool

Systems Engineer is both a single- and multi-user front end CASE tool operating under either Microsoft Windows or Presentation Manager. It provides support for the analysis and design stages of the life cycle, providing support for the techniques associated with these stages in LBMS' own methodology, Systems Engineering, in effect a version of LSDM designed to cover the whole life cycle.

LBMS refers to Systems Engineer as a groupware product. This concept can be summed up by the words 'cartoons to code'. The tool is based on Gupta Technology's client-server database product, SQLbase, and the facilities are arranged around a main menu of non-standard icons accessed by a cross mouse cursor. They include:

- data models
- dataflow diagrams
- dialogue design
- data inventory
- functional analysis
- project records
- help
- housekeeping
- sign-off.

Data models must be identified as current, required, third normal form, composite logical or user-defined. A data model can be split into one or more subsets, though this facility does not support many-to-many relationships, which have to be resolved immediately.

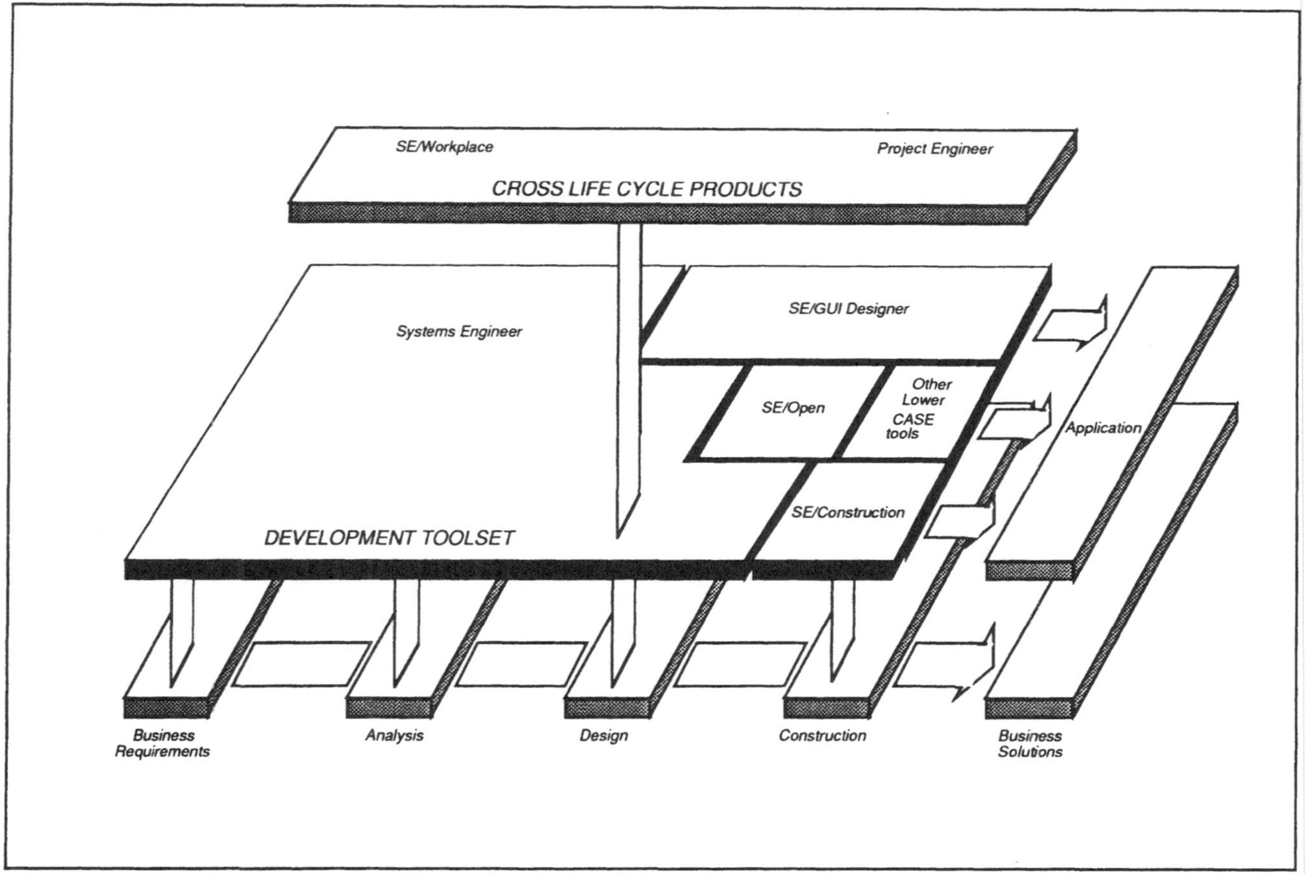

Figure 18.3 Systems Engineer Product Environment

The dataflow diagramming facility is based upon a top-down use of the technique. A specific project can have a number of DFD sets, where a set is based upon a number of decomposed DFDs.

Dialogue design is performed using menu control structure sets, menus and screen applications in conjunction with a design editor and screen map editor. The overall architecture of the system is represented as a framework of menus and transactions linked by control flows. A global definition picture is used to represent a dialogue sequence which has a specific function and is by default available anywhere in the menu structure. The transaction dialogues show the communication aspects of a transaction in terms of inputs, outputs and control passing.

The data inventory facility provides an ability to maintain information about data items and their associations.

The functional analysis facility concentrates on three design objects: functions, events and transactions. A function, in SE terms, is a grouping of closely related transactions or DFD processes, for which an overall description can be provided. An event is something that happens in the real world which acts as a trigger to processing. A transaction is a discrete unit of business processing. The user is supplied with a fill-in form to input details.

Project records contain the following applications:

- problems and requirements
- solutions
- general form types
- general forms
- general pictures
- general picture objects
- entity life history design
- batch suite design
- system structure design.

The first two are simple fill-in form facilities. The next four allow the user to define his own forms, pictures and objects, thus permitting extensibility of the tool. The last three are particular design editors.

There are two types of housekeeping applications: database control, which is used by a systems manager, and user maintenance, which maintains details about users.

5 Information Manager: the repository

This product is based on the old Meta Systems repository, which was portable across several platforms. The first version of Information Manager, running under MS/DOS and Windows, can be used in a LAN environment to support the integration and control of multiple Systems Engineer workgroups. The GUI allows easy manipulation and administration of repository data. Facilities include:

- security maintenance
- package definition (a grouping of meta-objects)
- check-in and check-out of packages
- reporting
- a browse editing facility.

The interface between Information Manager and Systems Engineer enforces access controls imposed by the Repository Administrator, such that only one SE database may hold update access rights to each package defined on the meta-bases, while several others may hold read-only copies of the package. Information Manager will also ensure consistency across design information from a number of SE databases so that, as workgroups share information, the organisation moves towards a more consistent view of this information. Information Manager will also work with Systems Engineer workgroups over WANs in the same way.

6 System generation via SE/Construction, SE/GUI and SE/OPEN

LBMS' approach to code generation is to supply two ways for users to meet their needs – SE/Construction and SE/GUI – plus an interface to other products – SE/OPEN. Figure 18.4 shows the two proprietary approaches to code generation.

SE/Construction (Applications Engineer) provides the capability to generate applications from information stored in Systems Engineer. Current target environments include:

- COBOL, IMS/DC and DB2 (the initial release)
- COBOL, CICS and DB2
- PL/1, IMS/DC and DB2
- COBOL, TPMS and IDMS/X.

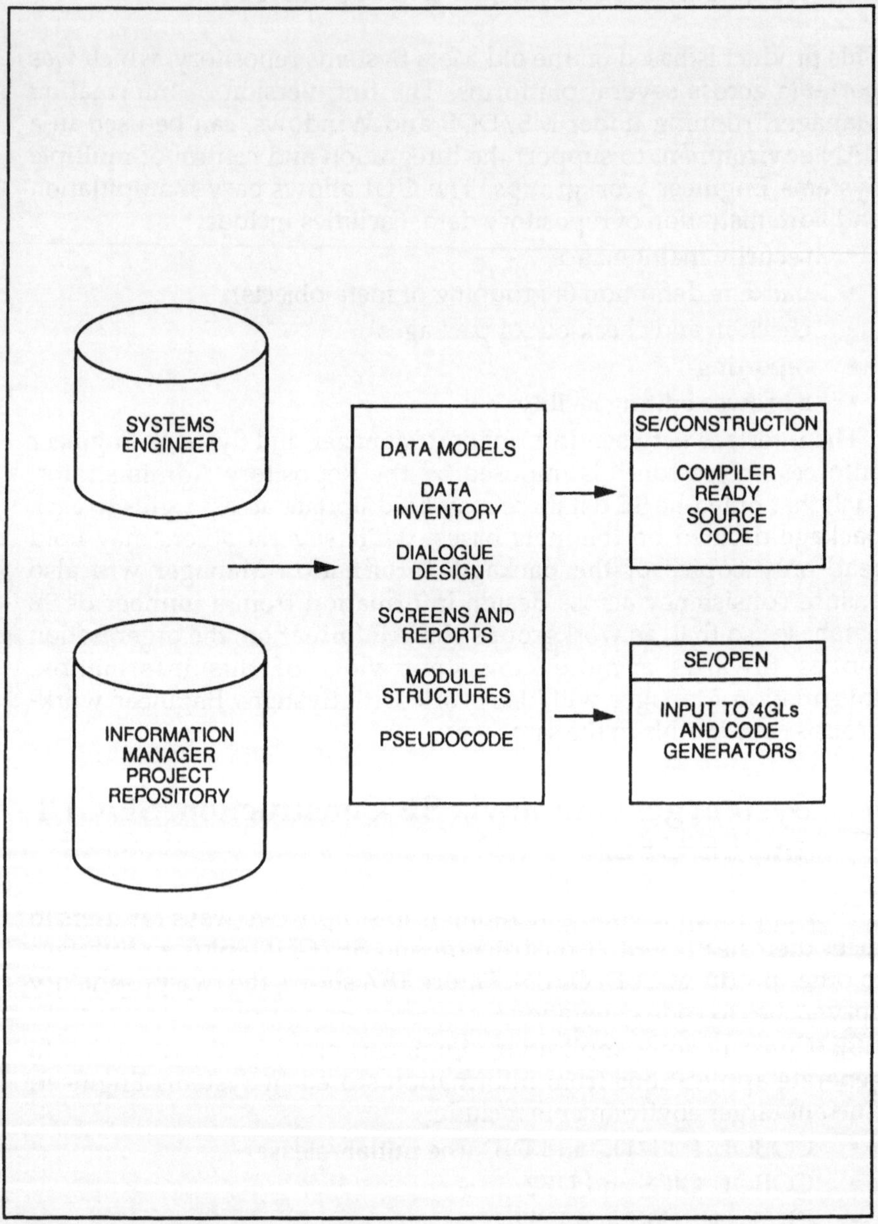

Figure 18.4 LBMS Code Generation

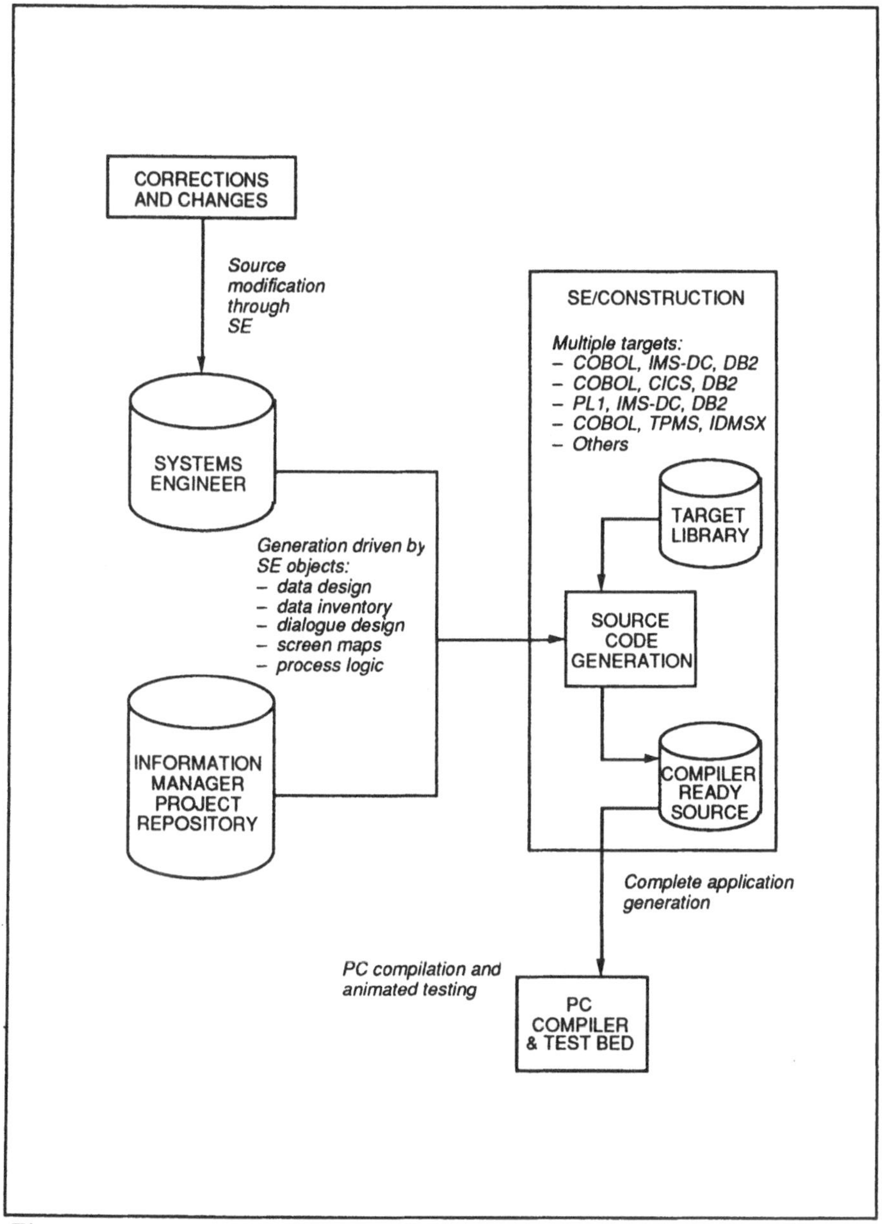

Figure 18.5 SE/Construction Environment

This product is based on technology derived from the acquisition of Michael Jackson Systems Ltd. The technology has also been used to provide ADA code generation for the real-time market. Figure 18.5 shows pictorially the SE/Construction environment and, in particular, the use of MicroFocus Workbench to provide a PC compiler and testbed.

SE/GUI provides a generator for GUI applications. The product takes an object oriented approach to the design of applications, and can convert a 3270-based application to a Windows-based one, thus providing a downsizing capability. SE/GUI consists of a dialogue editor, an image editor, and a font editor. Its inbuilt consistency checker verifies compliance with IBM's CUA standards. The product will generate applications for the following environments:

- MS/DOS, Microsoft Windows, C and Microsoft SQLServer
- OS/2, Presentation Manager, C, and DBM (supplied as part of OS/2 Extended Edition)
- OS/2, Presentation Manager and COBOL.

A future release will support applications running under OSF/MOTIF.

The third option, based on SE/OPEN, provides a CDIF-based interface to transfer information from Systems Engineer to back end CASE tools. At the time of writing, LBMS has delivered the following interfaces using SE/OPEN:

- UNiSYS LINC UP – providing an ability to generate LINC II screens and reports from Systems Engineer
- Intersolv APS BRIDGE – providing a link to Intersolv's APS product from Systems Engineer
- a link to Digital's CDD/Repository – allowing LBMS to be part of the COHESION platform
- a link to CA's IDMS, Version 12.

Other environments for which links are currently planned are :

- Pansophic's Telon
- ICL's DDS
- Ingres 4GL on ICL
- IBM's CSP.

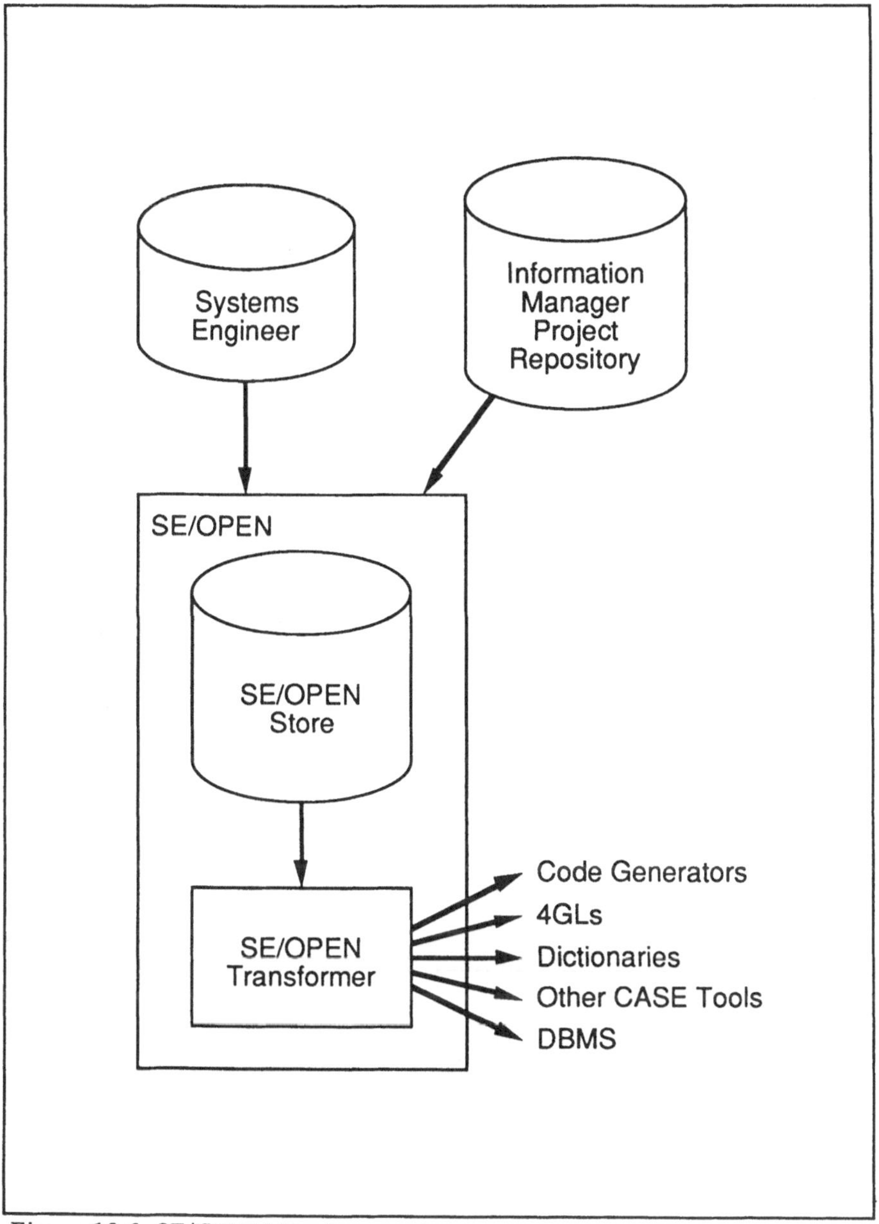

Figure 18.6 SE/OPEN Environment

7 REVENG: the reverse and re-engineering tool

The REVENG tool set is based on the single-user technology acquired from Meta Systems, and as such is centred around Information Manager. Figure 18.7 shows pictorially the components of that toolset.

The code analyzer is based on configurable parser technology that has already been adapted to a number of dialects of COBOL, FORTRAN and C. LBMS provides consultancy support to allow environments not already covered to be tailored and developed for individual customers. The results of the parsing are placed in Information Manager.

Work is under way to enhance REVENG to support dynamic analysis through the use of techniques for code instrumentation; this allows data to be captured, while the system is running, that models its dynamic behaviour. Dynamic analysis also allows the generation of test strategies for software maintenance.

At the time of writing, LBMS have stated that the REVENG technology is being actively phased out, in favour of a future standardisation around the multi-user Systems Engineer platform. Uncertainty remains (maybe deliberately encouraged?) about exactly what stage has been reached at any given moment.

8 LBMS Workplace: the tool integrator

LBMS Workplace integrates Systems Engineer or SSADM Engineer with other tools:

- user-selected desk top tools such as word processors, spread sheets and E-mail
- LBMS' own on-line method documentation and project management aids.

This linking is achieved through the use of Microsoft's Dynamic Data Exchange (DDE), which supports the maintenance of *hot links*, enabling one application or window to obtain data from another, in real time and on demand. The first release of Workplace provided DDE support for Systems Engineer and other elements of LBMS' Windows tool-set, and for other Windows applications such as Word for Windows, Ami Pro, Excel, Designer, Powerpoint, Microsoft Project for Windows, and Da Vinci E-Mail.

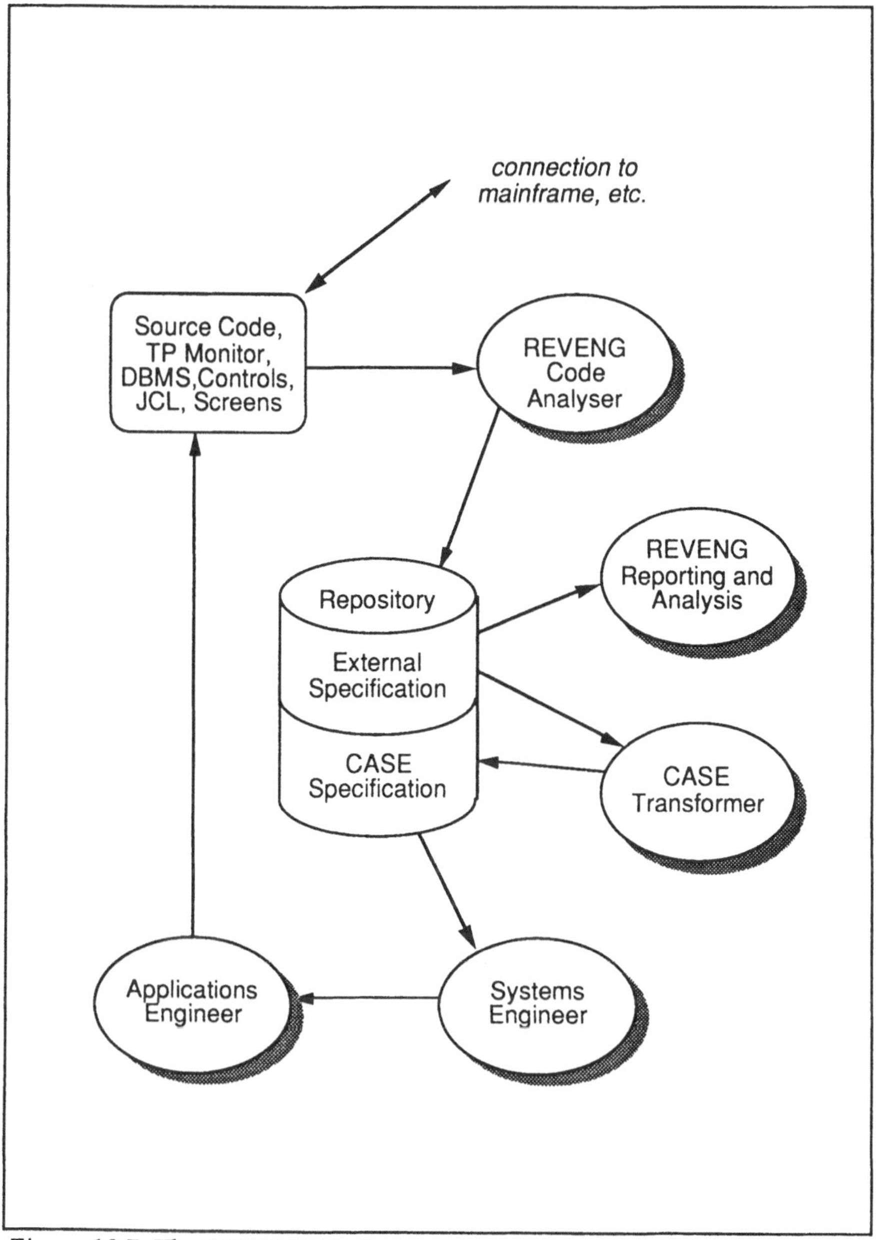

Figure 18.7 The REVENG Toolset

An important part of Workplace is a special query facility for Systems Engineer. This allows users to build their own queries on the underlying repository database. Once a query has been specified, other Window applications that support the *paste link* facility are able automatically to embed the special control information required to execute that query through DDE, retrieving the specified information when required.

Another way of hot linking applications will be Object Linking and Embedding (OLE), another Microsoft standard feature which requires Windows 3.1 for its operation. OLE support will be added to both Workplace and Systems Engineer now that Windows 3.1 is available, and will enable the embedding and linking of objects external to Systems Engineer into its database, as well as linking and embedding data and graphics from Systems Engineer in those external objects.

9 Project Engineer: the project management companion

Project Engineer is positioned by LBMS as a close companion for Microsoft Project for Windows. It is delivered with several standard template work breakdown structures (WBS) that can be used for a variety of projects using LBMS' methods. It supports the manipulation and customisation of these templates, which can then be loaded into Microsoft Project for Windows for further scheduling and planning.

The stated future direction for Project Engineer is to process project data from Project for Windows, building a database from which progress may be more closely tracked, and providing a framework of estimating factors.

10 Methods Engineer: the on-line hypertext methods support tool

Currently LBMS has released two versions of Methods Engineer: LSDM On-Line supports LSDM (Systems Engineering), and SSADM On-Line supports SSADM Version 4. The product is based on hypertext and hypergraphics technology, runs under Windows Version 3, and can be iconised so that it can be referred to when using Systems Engineer.

The information in Methods Engineer is stored in small chunks of text and graphics called frames. These are cross-referenced by a network of links, by which the user moves from frame to frame. The first thing seen by the user is an overview of the contents. A mouse pointer is used to select the subject of interest. A full frame of information on that topic is then displayed. Clicking on a phrase or picture causes a new frame of information to be displayed.

The information is arranged in two hierarchies for LSDM/Systems Engineering: one for the work breakdown structure (WBS) and the other for techniques. For SSADM, there are three hierarchies: one for the structural model, another for the techniques and a third for the glossary. These hierarchies are cross-referenced so that the user can move directly between them.

In the future, LBMS will integrate the LSDM/Systems Engineering version even more closely with Systems Engineer, so that the relevant method topics may be automatically looked up from the software context.

11 LBMS and other CASE strategies

LBMS is a member of IBM's Tool Vendor Assistance Program. Under this programme, LBMS is publicly committed to the following deliveries:

- an OS/2, Presentation Manager, CUA-compliant version of Systems Engineer that will be fully compatible with AD/Cycle Tools
- an External Source Format (ESF) interface to CSP and Synon/2 (the first version has already been released, supporting data object definition transfer)
- an online, interactive interface to IBM's Repository Manager (RM/MVS).

LBMS has CASE partnerships with both ICL and UNiSYS, with whom it is involved as a third party in the Integrated Information Environment Architecture.

LBMS is also active in the Open Standards community. It holds the vice chairmanship of the CDIF committee, and is fully committed to support the CDIF format once the standard has been formally approved. It is also involved in the development of IRDS standards.

In our view, LBMS is one CASE vendor that has recognised the importance of working with other products from other vendors, whilst at the same time providing through this interworking the capability of building a complete CASE environment.

12 Further reading

1 Gilpin, Mike (1991), *LBMS I-CASE: Overview and Statement of Direction*, LBMS, 23rd February 1991

2 Ormrod, Tracy-Anne (1991), *Doing CASE on Windows 3.0*, *.EXE* Magazine, Vol 6, Issue 1, June 1991

3 LBMS (undated), *Systems Engineer*, sales brochure

4 LBMS (1991), *Workplace Datasheet*, March 1991

5 LBMS (1991) *Application Engineer Datasheet*, February 1991

6 LBMS (1991), *SE/Open Datasheet*, June 1991

7 LBMS (undated), *LSDM On-Line*, sales brochure

8 LBMS (undated), *SSADM On-Line*, sales brochure

9 LBMS (1991), *An Overview of LBMS Strategic Planning*, May 1989

10 LBMS (1991) *The System Engineer Toolset*, sales brochure, October 1991

Chapter 19

Texas Instruments: Information Engineering Facility

1 Introduction

Texas Instruments (TI) is a high technology company with manufacturing facilities in 18 countries, and sales and marketing operations in more than 30 countries. TI develops, manufactures and markets semiconductors, defence electronics, software productivity tools, computer systems, industrial automation and control systems, electrical controls, metallurgical materials and consumer electronic products. The company headquarters are in Dallas.

The mission of TI's Information Technology Group is to provide cost-effective technologies that help their customers gain significant competitive advantage from the timely, effective use of information. Their strategy focuses on providing software, hardware and services to help their customers create information systems and to perform and manage business processes across their enterprises.

The Information Engineering Facility (IEF) was developed by TI in conjunction with James Martin Associates (JMA) during the middle 1980s. The product was launched in 1987. There are now some 450 users worldwide, with about 130 in Europe including 45 in the UK.

The IEF was one of the first life cycle CASE products on the market, supporting the system development life cycle from planning, through analysis and design, to construction and implementation. Initially the product was marketed by TI in the USA, and by JMA throughout the rest of the world. In 1990, TI purchased that part of JMA that dealt with the support and marketing of the IEF. A statement of direction for the product was published in 1991.[1]

2 Architecture of the IEF

Before going on to describe TI's strategy for the IEF, we will look at the underlying architecture of the product set (see figure 19.1). The IEF was designed from the start to support the techniques and deliverables of information engineering (IE). Inspired by James Martin's seminal vision of information systems engineered precisely to meet business requirements, the product transforms pictures and associated definitions into code and database definitions.

2.1 Information engineering (IE)

IE focuses on the business requirements of an organisation, and seeks to model how a business can best use information as a corporate resource. The life cycle therefore begins with information strategy planning (ISP), in which the business objectives and information needs are assessed for the whole, or a significant part, of an organisation. This produces a business model which represents data, activities and their interactions. Initially this model is at a conceptual level but as time passes and projects go through analysis and design to construction, so more detailed conceptual, logical and physical information is added to the business model. The IEF toolset provides the developer with three views of the business model, based on data, activities and their interaction. As IEF projects progress from ISP to business area analysis (BAA) and business system implementation (BSI), the model divides into business areas, which in turn divide into business systems. One project at the ISP stage may generate ten or more BSIs, all based on a single business model.

Throughout the stages of IE, the developers collectively access a single development database, known as the central encyclopædia. This provides a central repository for controlling a project and ensuring that all developers work from the same overall business model. This encyclopædia is located on a mainframe, initially on an MVS platform using DB2 with supporting ISPF interactive dialogs.

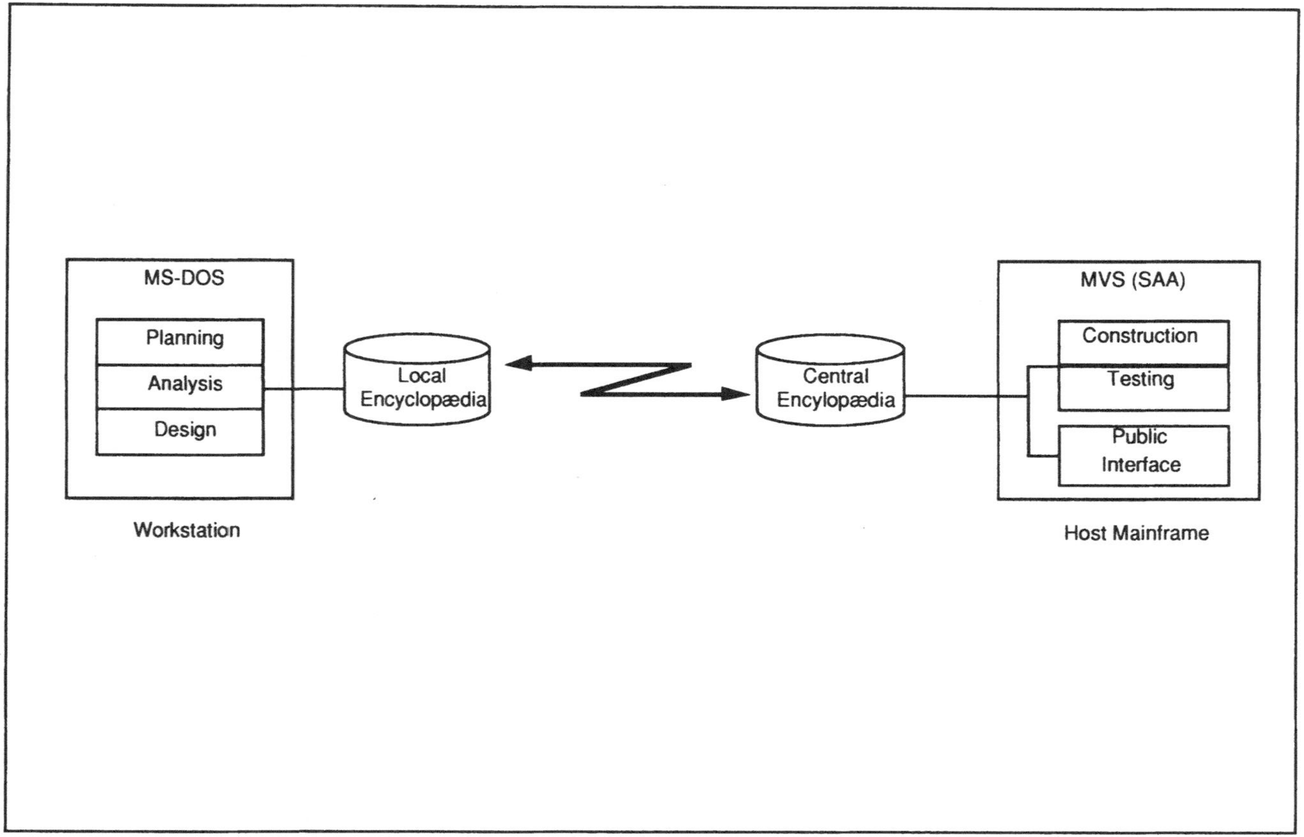

Figure 19.1 Current IEF Application Development Architecture

2.2 Encyclopædias

Each developer has their own local encyclopædia on their workstation, where they store the part of the business model that they are working on. The business model parts are checked in and out of the central encyclopædia using download and upload facilities. Each local encyclopædia provides a local reporting mechanism, and manages information at the workstation level.

Both the central and local encyclopædias are based on a governing meta model – the information model – which defines all the generic classes of objects in the IEF architecture and details their relationships to one another. This information model is not currently in the public domain, but figure 19.2 shows a schematic of the model indicating the basic concepts covered.

The IEF stores the definitions created through the pictures, rather than the pictures themselves. Diagrams are displayed at the workstations by recreating graphical symbols and relationships from the stored definitions. Early versions of the product could not accommodate user preferences for the positioning, shape and size of graphical objects, but current versions allow the user a degree of control over certain of the graphical elements, though less than most other CASE products on the marketplace – TI's belief is that the graphics' key purpose is to act as a window on the encyclopædia and, as such, the job is handled perfectly adequately.

2.3 IE support and the IEF

While the central encyclopædia gives the IEF its foundation, IE provides its structure. IE recognises seven formal stages in the system development life cycle:

- information strategy planning (ISP) – high level business requirements are identified, and a blueprint produced for developing a coherent set of systems
- business area analysis (BAA) – a defined business area is thoroughly examined and the business requirements are captured in detailed, documented diagrams
- business systems design (BSD) – those external aspects of the business system that directly concern the user, such as screens and commands, are designed

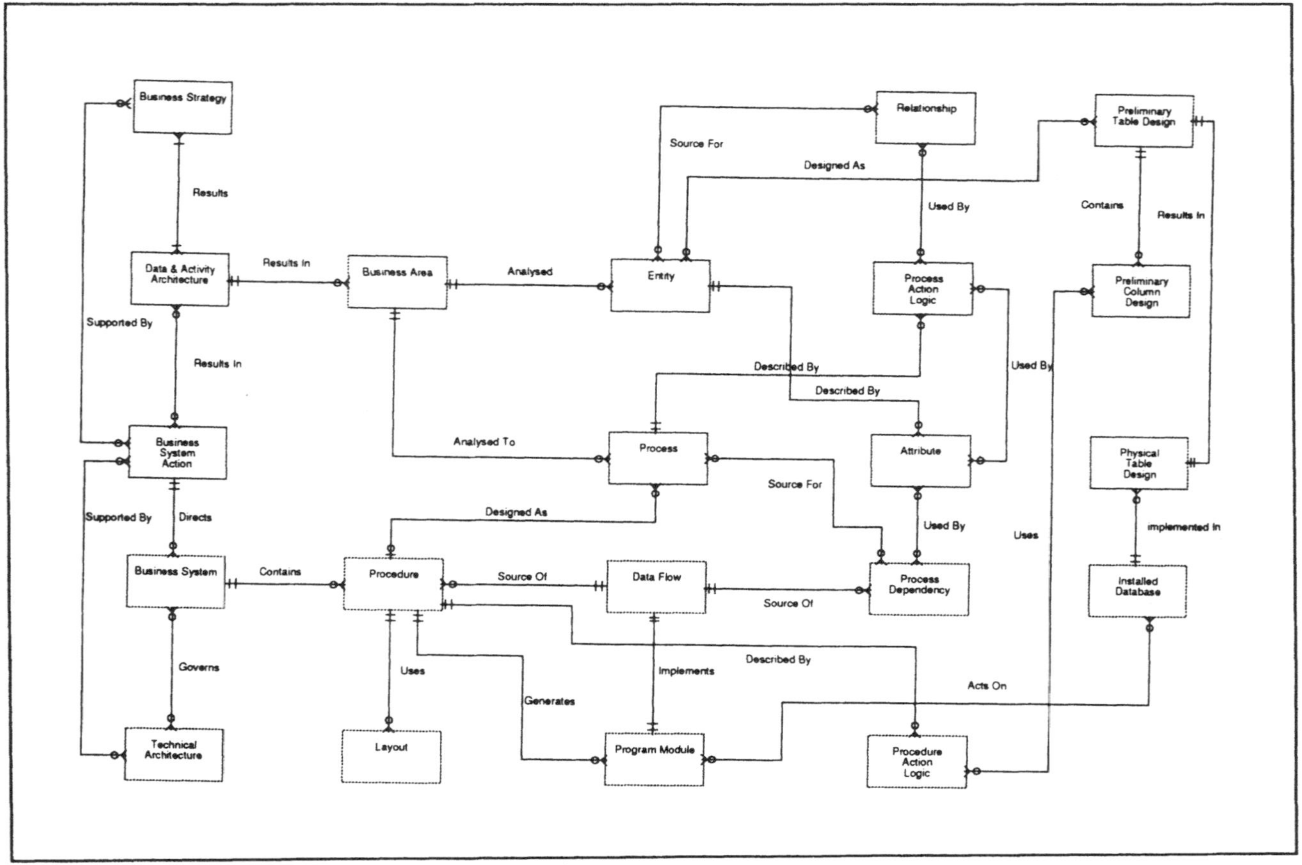

Figure 19.2 Schematic Information Model for the IEF

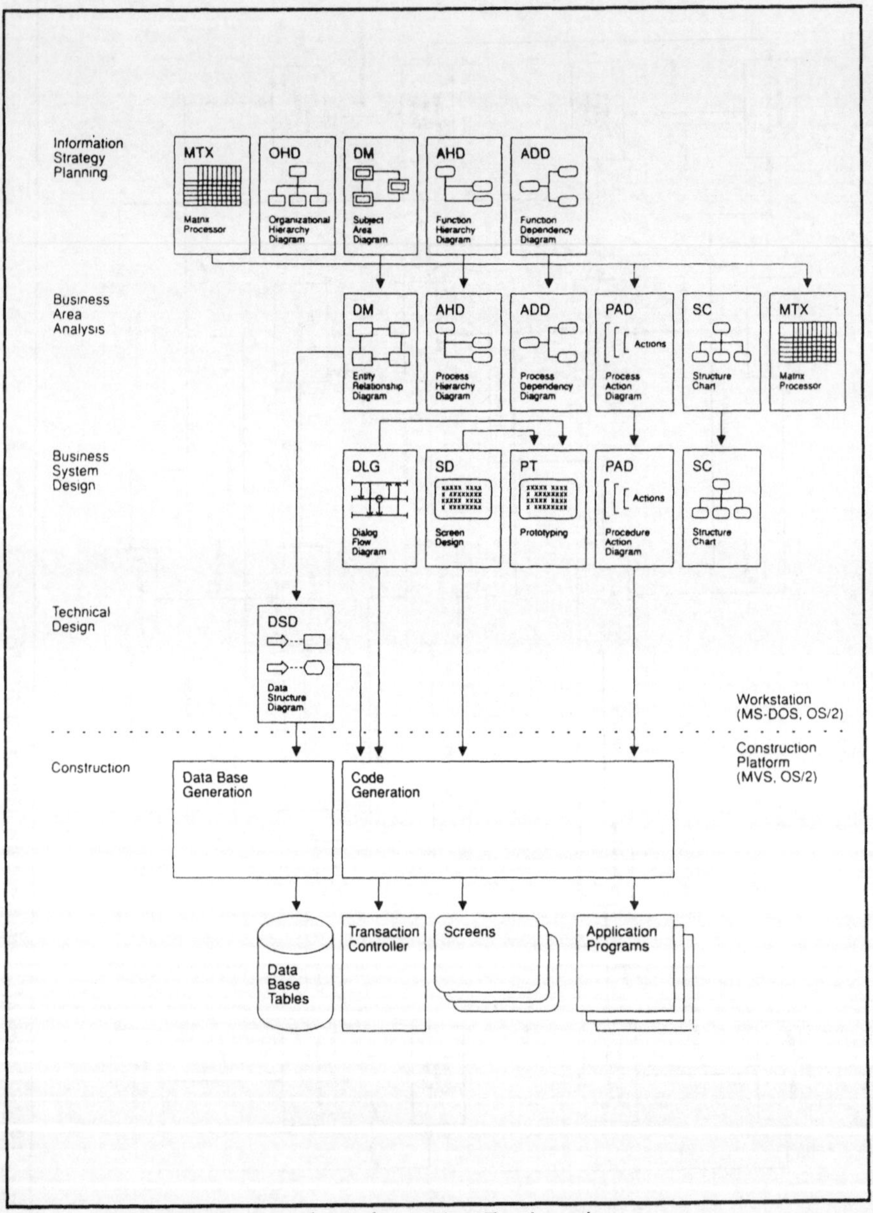

Figure 19.3 IEF Support for Information Engineering

- technical design (TD) – the model is tailored to achieve the most economical and efficient performance using a chosen technology
- construction – the system is coded and tested according to the detailed specifications produced during the previous stages
- transition – the database is loaded, new application systems are phased in, systems testing is completed and enhancements are added
- production – quality monitoring and performance measurement take place, as well as maintenance to fine tune the system.

2.4 IEF toolsets

The IEF provides automated support for IE through a set of tightly integrated tools, as shown in figure 19.3. The classic seven stages of IE tend to be rolled, in practice, into three development stages, with production as a fourth. The level of automation provided by the product allows BSD, TD, construction and transition to be handled within the scope of a single stage – the BSI, or business system implementation. The use of the IEF on RAD – rapid application development – projects allows even more compression of the conventional life cycle.

These tools are bundled together (see figure 19.4) into four workstation-based toolsets – planning, analysis, design and construction – some of which can be installed separately (the analysis toolset is always present).

2.4.1 Planning toolset

The planning toolset produces deliverables to support the ISP phase of IE. It provides the following tools to support the definition of the blueprints:

- data modelling – used to begin building the conceptual data model in the form of a subject area diagram and identifying a number of key entity types
- activity hierarchy – used to begin building the activity model in the form of a function hierarchy diagram

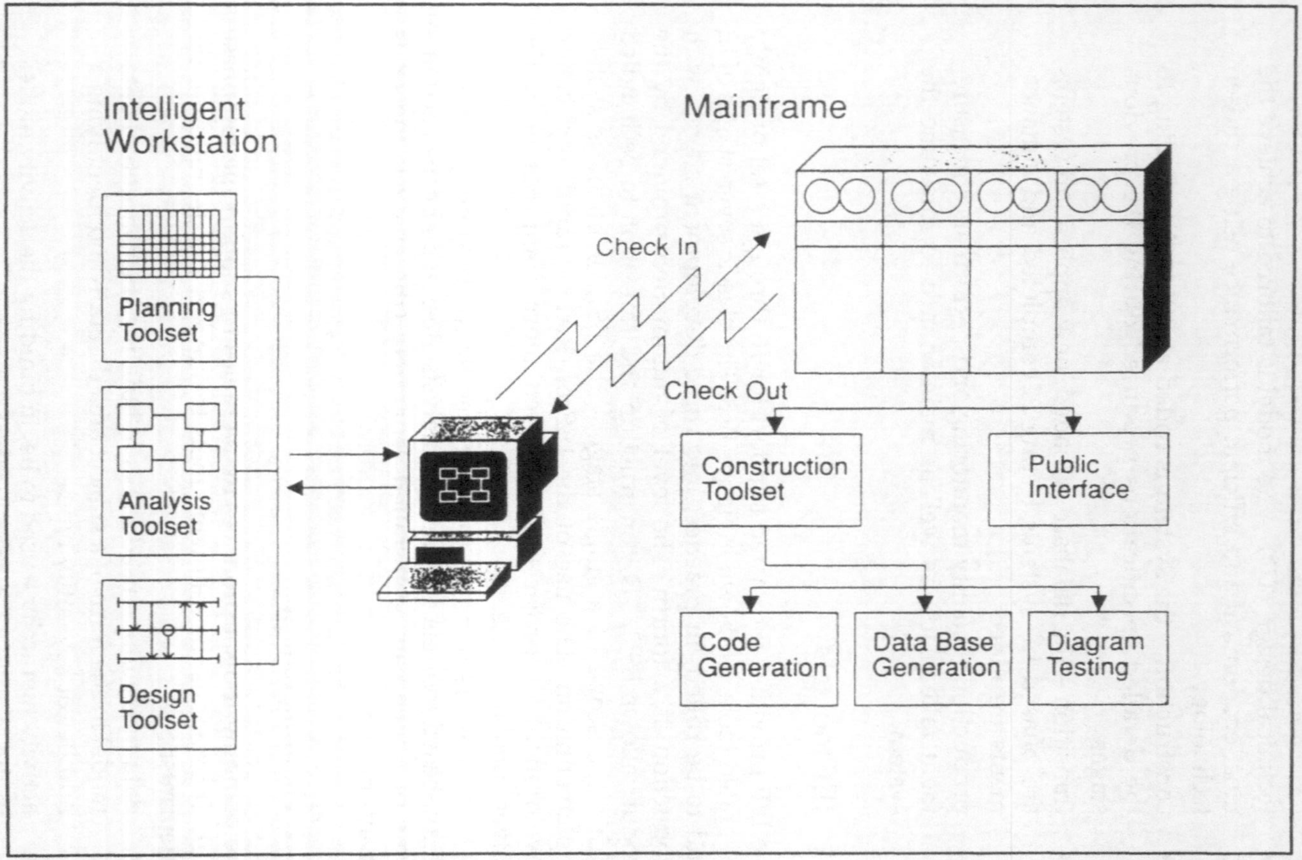

Figure 19.4 The Current IEF Toolset Environment

- activity dependency – used to validate the activity model in the form of function dependency diagrams
- organisational hierarchy – used in business planning
- matrix processor – used to conduct business planning, to begin building the interaction model in a business function/entity type matrix, and to define business areas in preparation for BAA
- matrix definition – used to define customised matrices.

2.4.2 Analysis toolset

The analysis toolset produces deliverables to support the BAA phase of IE. It supports business model development with the following tools:

- data modelling – used to continue building the conceptual data model in the form of an entity relationship diagram and a set of entity hierarchy diagrams (views of subtyping)
- activity hierarchy – used to continue building the activity model in the form of a function and process hierarchy diagram
- activity dependency – used to continue validating the activity model with a set of process dependency diagrams
- action diagram – used to continue to build the interaction model by means of process action diagrams
- structure chart – used to demonstrate how action diagrams interrelate from the top down
- action block usage – used to demonstrate how action diagrams interrelate from the bottom up
- matrix processor – used to continue building the interaction model using an elementary process/entity type matrix, and to define business systems in preparation for BSD
- matrix definition – used to define customised matrices
- business system definition – used as an alternative means of defining business systems in preparation for BSD.

2.4.3 Design toolset

The design toolset produces deliverables to support the BSD phase of IE. It supports design tasks with the following tools:

- dialog flow – used to continue refining the activity model in the form of a dialog flow diagram
- screen design – used to continue refining the interaction model by producing a set of user screens
- prototyping – used to simulate dialog flows between screens
- action diagram – used to continue refining the interaction model with a set of procedure action diagrams
- structure chart – used to demonstrate how action diagrams interrelate from the top down
- action block usage – used to demonstrate how action diagrams interrelate from the bottom up
- data structure – used to refine the physical data model in the form of a data structure diagram.

2.4.4 Construction toolset

The construction toolset produces code and database definitions for the chosen target environment from the information collected during the previous stages. It also provides an interactive diagram testing facility to check the system.

3 IEF key strategies

TI is committed to leading the industry in I-CASE technology. Therefore TI has stated that it will enable application developers to deliver systems with the focus on delivering business solutions, not on learning new languages and technologies.

There is also a commitment to standards and to ensuring that the IEF remains a flexible, open solution to information systems development for the widest possible community. This commitment has emphasised the urgency of moving the IEF from an IBM mainframe-only target, with DOS and (more recently) OS/2 development environments to a multi-platform product. This increases the potential marketplace for the product, whilst protecting it from the vagaries of IBM 's AD/Cycle tactics.

To achieve these goals, TI has adopted a number of key strategies. These are at the heart of TI's directions for IEF-generated application systems, and also for the development environment under which the IEF runs. The next sections look at these two environments and review the effect of the key strategies on them.

3.1 Research and development

In research and development terms, TI will continue to increase its investment in IEF research and development to gain advantage from new technology, both within IEF products and in IEF-generated applications. The first goal is to support industry standard graphical user interfaces (GUIs) and the client server model, for both IEF and IEF-generated applications.

There is a commitment to IBM: TI will continue its support for IBM users by supporting all SAA platforms, including AS/400, and the AIX platform. This is to be achieved by supporting the SAA, CUA and AD/Cycle programmes, both for the IEF and for IEF-generated applications.

There is an equal commitment to other vendors, and TI specifically names Digital, Tandem, Fujitsu and Hewlett Packard (HP). The strategy is to make the IEF and IEF-generated applications consistent with the various vendors' strategies for application development and application systems. Therefore, support will be provided for NAS and COHESION from Digital, and for NewWave and CASEdge from HP.

3.2 Open systems

TI will support the move towards open systems and will enable IEF products and generated applications to comply with open systems standards. TI's policy will involve:

- building interfaces to emerging repository products from IBM and Digital, and ensuring data sharing between the IEF and other vendors' CASE products
- utilising standard interfaces to enable data exchange between IEF-generated applications and other vendors'

products, or applications developed with other vendors' products
- seeking strategic alliances with key consulting groups to assure the widest acceptance of the use of I-CASE and IE, and ensuring that TI's customers have a choice of service suppliers
- promoting and supporting the use of IEF products with alternative methodologies, whether developed by TI's consulting alliance partners or by customers in-house.

3.3 Reusability and reuse

To support reusability and reuse, TI will enable IEF customers to exploit this key element of effective information management by:
- continuing to evolve the techniques of IE to exploit the benefits of object-orientation
- enabling reusable analysis and design knowledge bases, such as application and business templates and system stereotypes
- developing and enhancing methodology practices and procedures to maximise the potential of reusability.

3.4 Process management

To assist process management, TI will enhance the IEF's support for development planning and co-ordination, with particular emphasis on project management, change management and metrics.

3.5 'Globalness'

In terms of 'globalness' (sic), TI will provide support for multinational marketplaces. This will involve the provision of support services across the globe, either directly or through TI's strategic consultancy alliance partners. In addition the IEF and IEF-generated applications will be enabled to support local language requirements, including support for European, Arabic and Asian character sets.

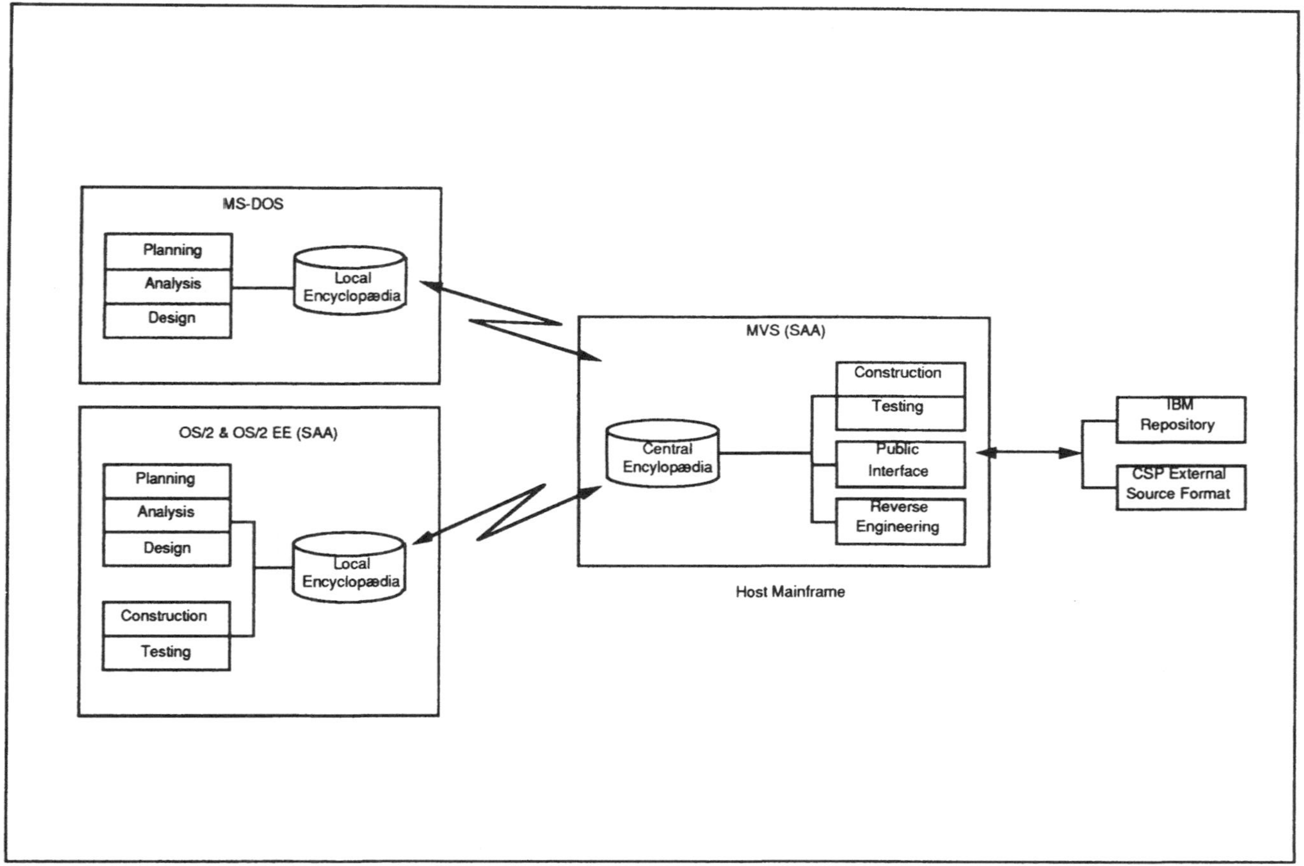

Figure 19.5 Planned Target Environments for the IEF

3.6 Customer focus

TI intends to demonstrate its customer focus by remaining responsive to the needs of IEF user groups, and by providing mechanisms to allow customer requirements to drive the future direction of the IEF.

4 IEF target environments

A target environment is a configuration of hardware and software under which IEF-generated applications can execute. Current target environments are:

- IBM MVS with DB2 under CICS or IMS/DC tp monitors, TSO or batch
- Digital VMS with either Rdb or Oracle DBMS
- TI System V with Oracle DBMS
- Fujitsu UTS/M with Oracle DBMS
- IBM OS/2 Extended Edition with DBM.

The short term plan (see figure 19.5) is to support the following additional platforms:

- the remaining IBM SAA environments, including:
  - VM with SQL/DS
  - OS/2EE with DBM under Presentation Manager
  - AS/400 CICS with OS/400 Database Manage
- Digital ACMS and DECforms under VMS
- Tandem Guardian fault tolerant environment with Tandem Pathway tp monitor and NonStop SQL DBMS
- Microsoft MS-DOS/Windows and OS/2 Standard Edition on IBM and IBM-compatible PCs and LAN servers
- IBM AIX on the RS/6000 range
- Hewlett Packard HP-UX on the HP 9000 range.

In the medium term (see figure 19.5), the focus is to be on supporting environments that conform to open systems standards including both the Unix International and Open Software Foundation versions of UNIX.

The support for DBMS products is to be increased, to include Ingres, Sybase SQLserver and Informix. Also, as standards for transaction processing emerge, TI will include support for suitable products.

The generated code currently supported is either COBOL or C, but TI plans to support the generation of ADA.

The IEF will support the design and construction of GUI applications for a variety of standard and de jure environments, including:

- Presentation Manager under OS/2 (both for standard and extended editions)
- Windows V3.x under MS-DOS
- MOTIF under UNIX and VMS
- Open Look under UNIX.

To aid the migration process to GUI applications, TI will provide 'screen-scrapping' facilities to re-engineer existing IEF applications. Looking beyond the current GUI human computer interfaces. TI will support image systems. This is to be achieved through a 'visualization' layer that will be added by the designer to standard IEF developed GUI applications. Initially support will be provided for applications under Presentation Manager and Windows.

The open systems strategy is exhibited in the target environment in terms of support for client server applications, not only through the planned 'screen scrapping' but also via the generation of cooperative processing applications. Initially, these will be limited to IBM's SAA platforms – OS/2 under Presentation Manager and MVS under CICS. In future, however, support will be available for heterogeneous environments comprising various GUIs, and operating systems. This will be achieved through the support of proprietary protocols, such as IBM's LU2 and LU6.2, and de jure protocols such as TCP/IP.

There is no stated intention to support the use of OSI protocols. This, in the view of the authors, will prove to be a weakness in the European market, where OSI has gained market acceptance. TI does, however, state that its goal is to ensure that both the IEF and IEF-generated systems conform with emerging industry and de facto standards, including compliance with the portability requirements published by key bodies, including the X/Open Portability Guide.

5 The IEF development environment

The development environment is the configuration of hardware and software used by the IEF during the development of application systems. In step with the TI's plans for a diversity of target environments, the range of IEF development environments is to be expanded.

Currently the IEF human computer interface (HCI) is mouse driven and, to an extent, windows-like. It runs under MS-DOS or, via Presentation Manager, under OS/2EE. The aim of TI is to make the IEF workstation HCI have the same look-and-feel in a number of GUI environments. To this end TI have adopted IBM's CUA graphical model as the standard for user interaction, though this commitment has not yet matured as a released product or even a scheduled release date. The following environments, having this common style of working, are to be supported:

- Windows V3.x under MS-DOS
- Presentation Manager under OS/2 (both standard and extended editions)
- MOTIF under UNIX (including AIX and HP-UX)
- MOTIF under VMS
- Open Look under UNIX.

To make the IEF attractive to a wider range of organisations, TI plans to make the central encyclopædia products available on a range of different platforms beyond the original MVS/DB2 implementation. This will include both versions of OS/2 using DBM, VMS and UNIX using Oracle DBMS and Tandem using NonStop SQL. The central encyclopædias (now called 'host encyclopædias') will be accessible across a LAN from any IEF workstation, and will support communication between multiple host encyclopædias (see figure 19.6).

The key strategic direction of open systems affects the IEF development environment through the need to support standard and vendor programs. TI is an active participant in the following standard and vendor programs:

- IBM's AD/Cycle
- Digital's COHESION
- ANSI SQL (X3H2)
- ANSI IRDS (X3H4)
- ANSI C++ (X3J16)

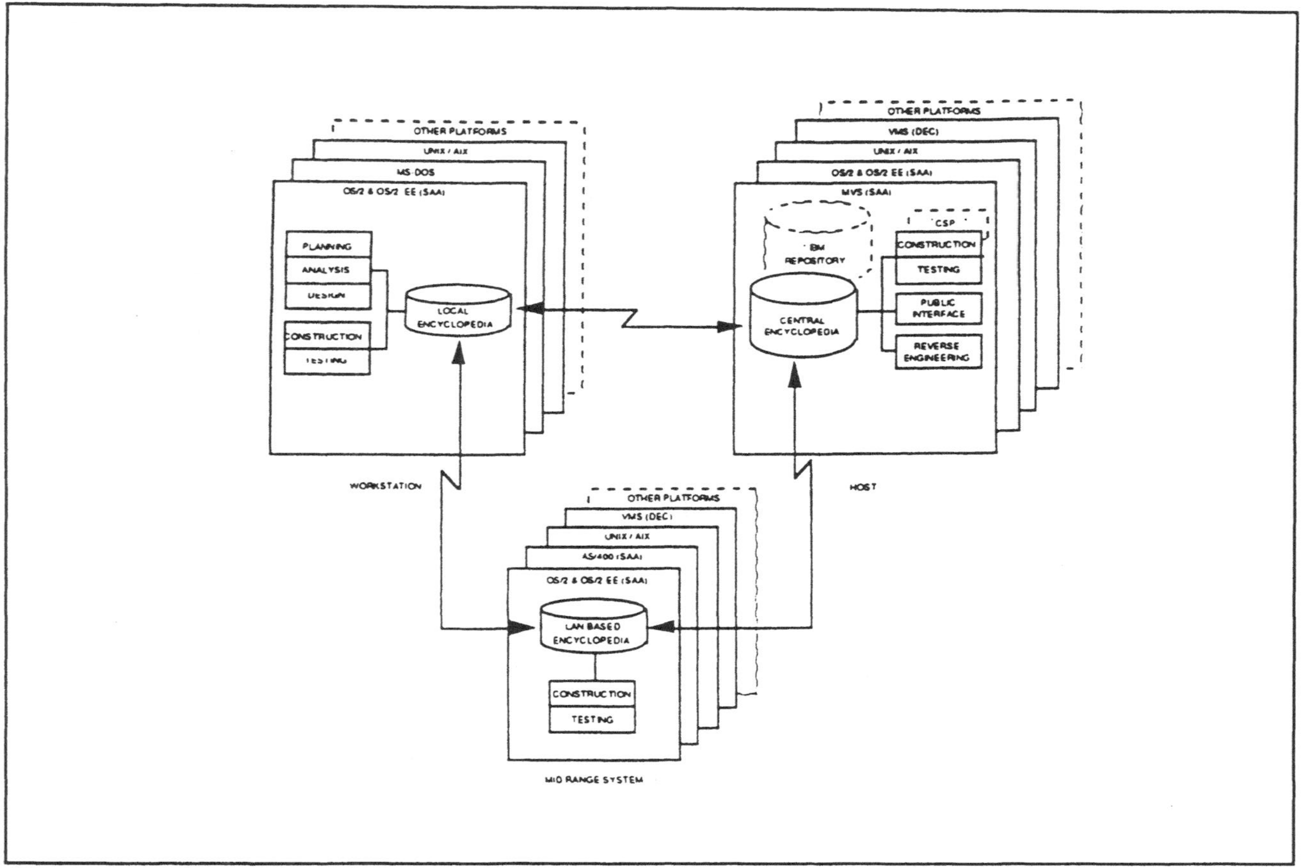

Figure 19.6 The IEF Planned Three Level Architecture

- UNIX International
- Open Software Foundation
- X/Open
- IEEE standard reference model for computer systems tool interconnection (P1175)
- International Function Point User Group
- Object Management Group
- Database Systems Study Group
- US National Institute of Standards and Technology (NIST) integrated software engineering environment (ISEE).

In terms of AD/Cycle, TI is working towards making the two information models closely compatible. TI's view is that the IBM information model is as yet inadequate and incomplete but that, as more is made available, TI will enable close integration between the IEF and the Repository Manager product. TI has already developed a facility to export information from the IEF using IBM's ESF standard.

TI is working closely with Digital to ensure that the IEF can exchange information with CDD/Repository using Digital's proprietary integration mechanisms.

The open architecture also means that the IEF requires two-way interfaces to other vendors' CASE products. The open architecture element of the IEF toolset forms part of the host encyclopædia, and enables information to be exported from or imported to the encyclopædia. As part of this programme, TI has committed to using Software One's Exchange product to provide two-way interfaces. TI is also looking at the use of the emerging CDIF standards as an alternative approach.

The idea that information might be imported into the IEF from other products has been a radical one for its developers to grasp, since the methodology rules and checks within the product set are so rigorous that nothing inconsistent can be entered in the normal course of an IEF project. We await the results with interest.

6 Conclusions

TI's IEF strategy seems to be concerned with keeping the product at the forefront of the integrated CASE world, whilst all around the marketplace is moving into the world of open systems. The IEF has always suffered in the past from its poor ability to interface and work with other products. The concept of I-CASE may be the ideal but it is one that, in practice, will only appeal to a real 'greenfield' user. The major theme behind TI's strategy appears to be aimed at opening up the product set to work on different platforms under different GUIs, whilst at the same time being able to exchange information with the 'big boys' (IBM and Digital). The problems of configuration management have only just begun to be addressed, and this may well limit the effectiveness of the checking in and out procedures.

TI's IEF statement of direction provides, for current users of the product, the basic statements necessary to reassure them that they will be able to use the product set beneficially in the open world of the 1990s. As with so may other vendors, TI has to make good on its intentions. The authors believe that this may be achieved within the wider context of a fully distributed development and live system environment, which TI hopes will account very soon for a substantial proportion – conservatively 10% – of its total worldwide revenue. That is very big business indeed. Watch this space!

7 References

1 Texas Instruments (1991), *Information Engineering Facility (IEF) – A Statement of Direction*

Appendix

Total Quality Management (TQM)

1 Introduction

As Richard E. Zultner has pointed out,[1] 'Every software organisation wants to be world class. Every software professional wants to be the best he or she can be. Yet very few users would nominate their IT organisation for world-class honours. ... Quality is a major problem for software.' A significant part of this Appendix is derived from Zultner's comments.

In Japan, there are many examples of companies using total quality management (TQM) to turn themselves around or deal with corporate crisis, and there is a widespread awareness that quality leads to productivity and profitability. Quality and productivity have been seen as two side of the same coin.[2]

TQM is an idea whose time has come, according to Jon Choppin.[3] 'The political ramifications of the decisions taken by the leaders of the European Community at Maastricht in December 1991 will continue to reverberate for some time. ... The nub of the debate ... was about the control of industrial change. ... The movement towards total quality management both in large and small companies represents the type of change with which the Euro-planners were concerned. ... The emphasis of the social charter was on the rights of workers (and) the energy to develop the other areas of TQM is provided by maximising the potential human resources of the company.' Choppin used two diagrams (see figure A.1) to explain the fundamentals of TQM. The three fundamental components of TQM are unit optimisation, horizontal integration and vertical alignment.

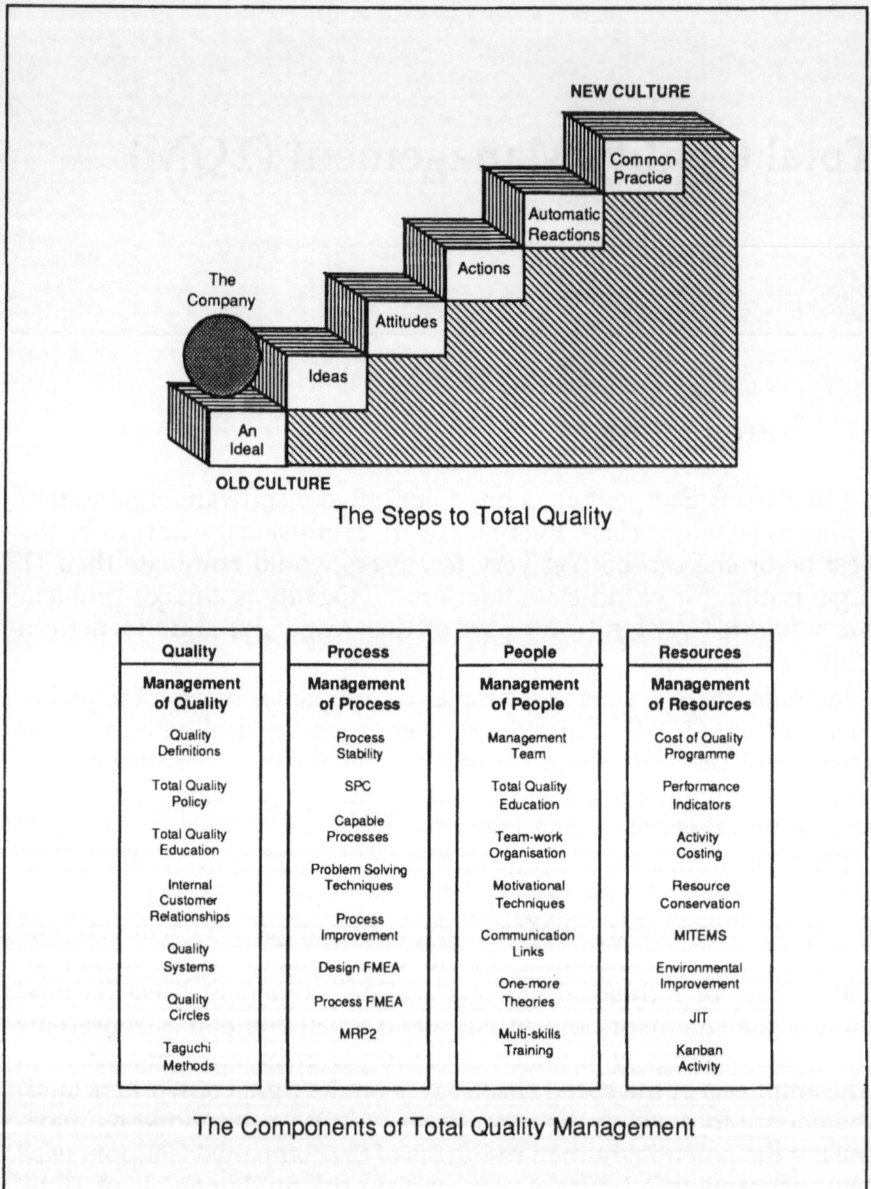

Figure A.1 Steps to TQM and Components of TQM (after Choppin)

2 Unit optimisation

All individuals and groups of individuals of the organisation are responsible for stabilising and improving their own work. This reduces variability and provides a baseline for later improvement efforts. It requires knowledge of how to stabilise and improve work, and is carried in three steps: learn-do-check-act, kaizen (continuous improvement) and statistical process control. Much of this process depends on recurring cycles – see figure A.2.

2.1 Learn-do-check-act

A solid knowledge of software engineering is absolutely essential to the development of high quality software. This requires an extensive training program, and also that the knowledge is applied, the results studied, and action taken to build upon the lessons learned. This is the first step of TQM, and is called the learn-do-check-act (LDCA) cycle.

2.2 Kaizen (continuous improvement)

The next step is kaizen – continuous improvement – and consists of standardisation, incremental improvement and innovation.

Standardisation follows the standardise-do-check-act (SDCA) cycle. This reduces initial variability and locks in later gains.

Incremental improvement uses the plan-do-check-act (PDCA) cycle, which is used by everybody every week. Every normative task is planned, performed according to plan, and studied for lessons learned; then appropriate action is taken so that further improvement or standardisation can maintain the gains. The improvement effort is documented in a quality improvement story, which is presented to management.

Innovation is accomplished by breaking bottlenecks. The key is to consider what to change, what to change to, and how to change, and then to dissolve the barriers limiting you to today's results. Kaizen stresses data-driven improvements. Decisions are based on data, not experience, education or gut-feel. Thus metrics become essential.

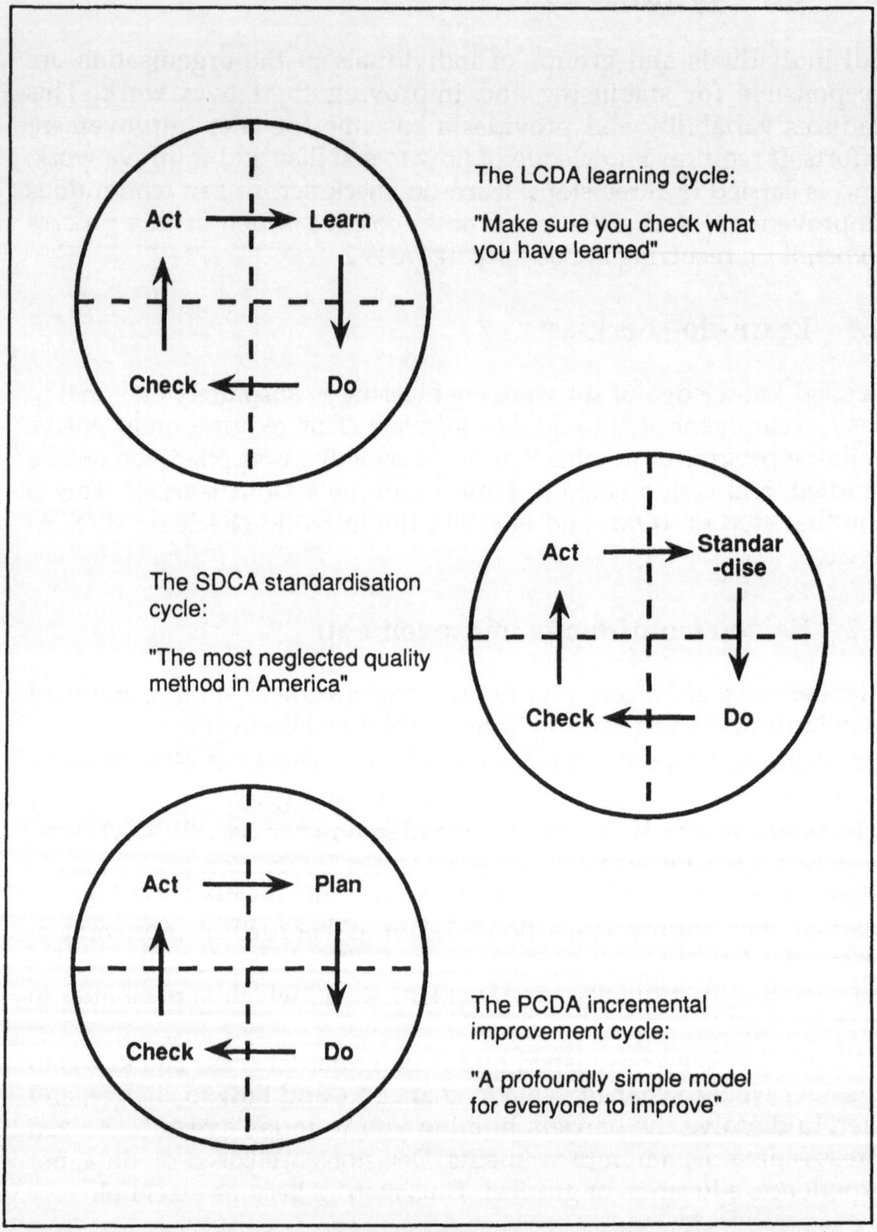

Figure A.2 The LDCA, SDCA and PDCA Cycles (Zultner's comments)

2.3 Statistical process control (SPC)

Standardisation is a prerequisite for meaningful metrics. To work with data intelligently, statistical process control is essential, as Zultner has noted:[1] 'SPC has a 50-year track record in manufacturing and works fine for software, too. At NEC IC Micon, several mathematical models have been developed from equations in U.S. software engineering texts and refined into accurate estimating tools through the patient accumulation of data from *all* ongoing projects. Today, everyone collects data. Everyone analyses data statistically (everyone is trained to do so). Everyone works with data (including management). SPC is the most basic quality system.'

3 Horizontal integration

All departments and project teams must concentrate on satisfying the user together across the functions of the organisation. Two fundamental concerns are efficiently managing these cross functional development activities and finding and focusing on the most valued user requirements. To tackle these, project management and quality function deployment are required.

Project management plans and controls resources, time and costs. To plan a project properly requires project management tools.

Quality function deployment (QFD) 'provides the focus on high-value requirements for effective software engineering',[1] and is illustrated in figure A.3. QFD tackles this both by improving the product and by improving the process. Comprehensive quality deployment addresses quality in terms of what the user values, technology, innovation and reliability. Functional task deployment is used to build quality activities into the software life cycle.

4 Vertical alignment

All managers and professionals understand and contribute to the few crucial goals (hoshins) of the organisation. This approach, developed in Japan deploys the strategic vision of the company to everyone, breaking it down into just a few goals, the managing director's hoshins. Each area reporting to the managing director breaks these

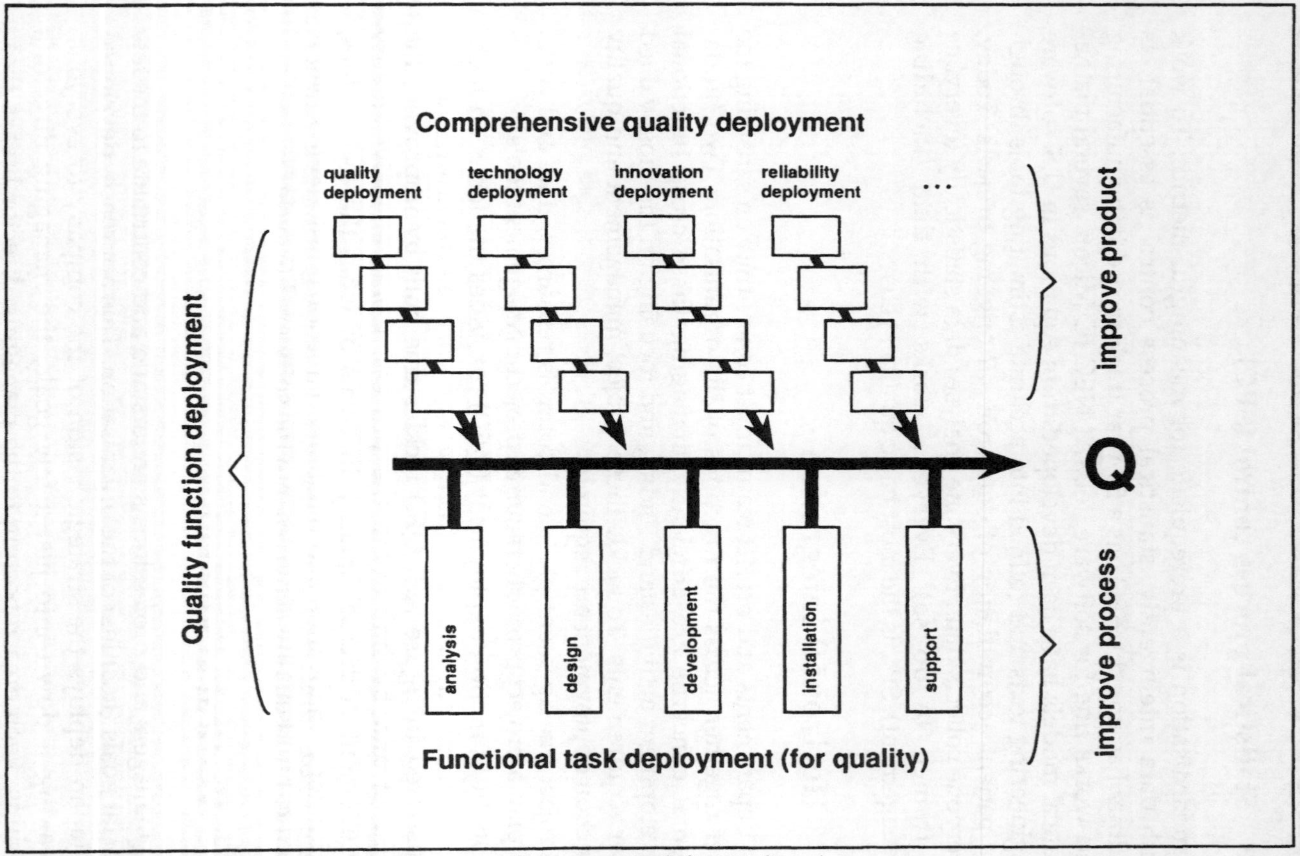

Figure A.3 Quality Function Deployment Components (after Zultner)

down into its own hoshins. This continues until everyone knows the vision and what they or their unit will do to contribute to that vision.

5 The quality crisis and Juran's trilogy

There is a quality crisis, and this will not go away in the foreseeable future. Competition based on quality will intensify, and the traditional ways in which organisations have tackled crises are not adequate to handle the quality crisis. A new course must be charted.

According to Dr. J.M. Juran, an organisation that wants to chart a new course in managing quality should create an 'all-pervasive unity' so that everyone will know the new direction, and will be stimulated to follow it. Juran's trilogy consists of three basic quality-oriented processes.

- Quality planning prepares and organisation to meet its quality goals, and results in a process capable of meeting these goals under operating conditions.
- Quality control ensures that quality goals are met during operations, and results in the conduct of operations in accordance with the quality plan.
- Quality improvement breaks through to unprecedented levels of performance, resulting in the conduct of operations at levels of quality distinctly superior to previous planned performance.

6 References

1 Zultner, Richard E. (1990), *Software Total Quality Management (TQM): What Does It Take To Be World Class?*, American Programmer, November 1990.

- Note: Much of Zultner's analysis was based on research done by GOAL/QPC, Kaizen Institute of America, Joiner Associates, the American Supplier Institute and 'other U.S. quality research organisations'.

2 Quoted from Lee Iacocca, ex-head of General Motors (exact source and date unknown)

3 Choppin, Jon (1992), *Striking back with TQM*, Management Consultancy, March 1992

Index

S

T

U

V